Croydon Libraries

You are welcome to borrow this book for up to 28 days. If you do not return or renew it by the latest date stamped below you will be asked to pay overdue charges. You may renew books in person, by phone or via the Council's website www.croydon.gov.uk

SHIRLEY LIBRARY
020 8777 7650

UNIT SHR 4/06		
	2 3 MAY 2014	
	27 SEP 2014	
1 4 DEC 2006	27 OCT 2015	
1 1 JAN 2007	2 1 NOV 2015	
1 3 MAY 2010		
1 0 MAY 2011		
2 9 DEC 2011		
1 8 APR 2012		

CROYDON COUNCIL
Cleaner, Safer & Greener

BAe 146/RJ
BRITAIN'S LAST AIRLINER

STEPHEN SKINNER

TEMPUS

By the same author, published by Tempus:
BAC One-Eleven – The Whole Story
Marshall of Cambridge

Frontispiece: G-LUXE (E3001) overflying Hatfield. (BAE Systems)

First published 2005

Tempus Publishing Limited
The Mill, Brimscombe Port,
Stroud, Gloucestershire, GL5 2QG
www.tempus-publishing.com

© Stephen Skinner, 2005

The right of Stephen Skinner to be identified as the Author of this work has been asserted in accordance with the Copyrights, Designs and Patents Act 1988.

All rights reserved. No part of this book may be reprinted or reproduced or utilised in any form or by any electronic, mechanical or other means, now known or hereafter invented, including photocopying and recording, or in any information storage or retrieval system, without the permission in writing from the Publishers.

British Library Cataloguing in Publication Data.
A catalogue record for this book is available from the British Library.

ISBN 0 7524 3562 0

Typesetting and origination by Tempus Publishing Limited
Printed in Great Britain

CROYDON LIBRARIES	
UNI	
02037782	
Askews	15-Mar-2006
629.133 SKI TRA	£19.99

Contents

	Acknowledgements	6
	Preface	7
1	The Delivery of the Final British-built Airliner	9
2	Initial Concepts at Hatfield	11
3	Start-stop-start – A Protracted Birth	15
4	A Technical Description	25
5	First Flight and Certification	37
6	The Stretched 146-300 Series	51
7	The 146 in Service	59
8	Royal and VIP Connections	85
9	Quiet Trading	95
10	Military Versions	105
11	In and out of the City	113
12	From Hatfield to Woodford	120
13	Rebranding – 146 to Regional Jet	125
14	The RJ in Service	131
15	Accidents and Incidents	141
16	The RJX and the End of Production	150
17	Managing and Developing the Assets	161
18	Conclusion	168

Appendices

1	Flying the BAe 146-300	171
2	BAe 146/RJ/RJX Production List	174
3	Chronology of the BAe146/RJ/RJX	186
4	Aircraft Data Aircraft Data BAe 146 and RJ	188
	Bibliography	190
	Index	191

Acknowledgements

One of the joys in researching and writing a book is the opportunity it affords to meet and discuss the aircraft with many different people who were involved with it. This book has been no exception and I am indebted to many who gave so generously of their time.

From the many former Hatfield people who were involved in the genesis and development of the 146 I have to thank Alan Blythe, Brian Botting, Maurice James, Johnnie Johnstone, John Loader, John Martin, Roger de Mercado, John Payne and Ken Pye, who discussed the aircraft with me and several of whom kindly lent me material.

I also owe a debt of gratitude to people from the recent years of the aircraft at Woodford, to Ken Haynes who answered many detailed questions, and especially to Derek Ferguson, who gave me access to his excellent private 146/RJ photo collection and arranged a visit for me to the Avro RJX at Woodford.

During a visit to RAF Northolt in April 2004 crew members of No.32 (The Royal) Squadron explained the 146's role with the Squadron to me in detail and gave me the opportunity to see over one of their immaculate aircraft.

Barry Guess and Mike Fielding of BAE Systems Farnborough were also very helpful in providing photos, as was Ian Lowe at Woodford and Terry Mitchinson, Editor of the *Welwyn & Hatfield Times*. Likewise Brian Riddle of the RAeS Library at Hamilton Place, who took some pains to unearth a good deal of pertinent material for me.

David Dorman, Head of Sales & Marketing BAE Systems Regional Aircraft, was very helpful in many ways. He arranged for me to visit G-LUXE at Cranfield and to meet Steve Doughty, Vice-President Sales & Marketing at Bishop Square, Hatfield, to discover more about the aircraft's past and especially its present and future.

My researches also drew me to the 146/RJ Test Pilots, Peter Sedgwick and Dan Gurney, who drew on their extensive experiences of testing and flying the aircraft around the world. Dan also gave me access to a large range of photos and additional materials. Chris Grainger, a CityJet 146 pilot, generously contributed the piece on flying into London City Airport.

My good friend and very skilled designer Rolando Ugolini (rolando@macace.net) versioned photos of early proposals and drew illustrations of the unflown military examples. I also acknowledge my gratitude to another friend, Derek Williams, who cast his gimlet editorial eye over the text.

I should like to thank *Flight International* for allowing me to reproduce the article on 'Flying the BAe 146-300' as Appendix 1.

My final tribute is to my wife, Jane, who not only gave me her steady encouragement but also provided editorial support throughout the writing of this book.

Preface

The story of the BAe 146/RJ attracted me for two reasons: it is Britain's best-selling jet airliner and also the country's last. So ends a tradition of continuous jet airliner manufacture, which began most illustriously with the de Havilland Comet in 1949.

Britain's final airliner type has a strong family connection with the de Havilland Comet as it was designed at the former de Havilland factory at Hatfield in the early 1970s, which had become part of Hawker Siddeley Aviation only a decade before.

By the time the 146 first flew in 1981, the Hatfield factory was part of the nationalised British Aerospace and in the following decade production of the 146 passed to the former Avro factory in Woodford while the Hatfield site closed. Like de Havilland, Avro is a famous name in the chronicles of British aviation, and produced many significant aircraft, particularly bombers such as the wartime Lancaster and Cold War Vulcan 'V' bomber.

Though development and production ceased in 2002 with the recent extension of the aircraft's life to 80,000 cycles, the 146/RJ will continue to represent the British Civil Aircraft industry as it quietly graces the skies.

1

The Delivery of the Final British-built Airliner

The final British-built airliner, an Avro RJ85 OH-SAP, was delivered on 26 November 2003 to the Finnish airline Blue 1, bringing to a close twenty-two years of BAe 146/RJ production. It also brought to a close a history and tradition of airliner manufacture in Britain, which began with the de Havilland 16 in 1919 and continued with many other firms throughout the ensuing eighty-four years. The BAe 146/RJ was initially assembled at the BAe plant at Hatfield, Hertfordshire, and a second production line was initiated in 1988 at Woodford in Cheshire. With the closure of Hatfield in 1993, production was concentrated at Woodford. This aircraft was the last of Blue 1's lease contract for four aircraft: two RJ100s and two RJ85s. It had previously made the final first flight of a British-built airliner from Woodford on 24 April 2002 registered as G-CBMH. It was later stored at Filton and was customised for Blue 1 by Flybe Aviation Services at Exeter.

On its delivery flight it overflew the BAE Systems facilities at Weybridge and then Hatfield where the 146 made its maiden flight and where a total of 166 BAe 146s were built. Many will recall the famous airliners that were conceived at Weybridge and Hatfield: Viscount and Comet, to mention just two. OH-SAP then landed at Woodford, the former 146/RJ assembly centre, and then took off for Prestwick, Scotland, the Headquarters of BAE Systems Regional Aircraft.

At Woodford OH-SAP posed in company with the prototype BAe 146 G-LUXE – the first and last BAe 146/RJs together. G-LUXE was on flight trials after a lengthy conversion at Woodford to become a meteorological research aircraft for the Facility for Airborne Atmospheric Measurements.

Between 1981 and 2003, a total of 394 BAe 146s, RJs and RJXs were built and in all 390 delivered to airlines and other customers. The odd ones out were the 146–200 prototype, which was scrapped at Filton in 1995, and the three

OH-SAP (E2394), the last BAe 146/RJ airline delivery, together with the first 146, now G-LUXE (E3001), at Woodford on 25 November 2003. (Ian Lowe)

RJXs; of which one is still at Woodford, another is at the Manchester Airport Museum and the third example was dismantled.

The BAe 146/Avro RJ proved to be Britain's most successful *jet* airliner, though the turbo-prop Vickers Viscount outsold it with its total sales of 438 and the Woodford-built Hawker Siddeley (Avro) 748 turbo-prop almost equalled it with 380 deliveries. The end of RJ production had originally been announced on 27 November 2001 when BAE Systems Chief Executive John Weston stated that the regional jet business was no longer viable in the post-September 11 environment. However, as BAE Asset Management has a substantial number of 146/RJs in its jet leasing portfolio its involvement with the type will continue for many years.

Henceforward BAE Systems will not build civil aircraft but will manufacture military aircraft and major components for airliners. And via Airbus, in which BAE Systems has a 20 per cent holding, the firm still designs and builds all the wings for the Airbus airliners and A400M military transport at the Filton and Broughton (Chester) Airbus UK plants.

2

Initial Concepts at Hatfield

In the late 1950s de Havilland was a powerful, independent, international organisation with airframe, engine and guided missile interests based at Hatfield in Hertfordshire. The firm had built its reputation on many famous aircraft such as the Tiger Moth, Mosquito, Vampire and, of course, the ill-fated Comet. As the decade came to an end, the firm was still building Comet 4s for world airlines and the first metal was being cut on the de Havilland Trident for which BEA had placed a large order.

Simultaneously, project work was taking place in 1959 on a 'Dakota replacement'. De Havilland had much expertise with small airliners and wanted to build on their experience with the pre-war Dragon Rapide and post-war Dove and Heron. Whereas the Trident was numbered the DH121, this project was the DH123 with a high wing, two DH Gnome turbo-props and capacity for thirty-two-forty passengers. In fact, it was not dissimilar to the already existing Fokker Friendship, and British-built Avro 748 and Handley Page Herald. One year later the project engineers under Derek Brown had re-thought the specification and made substantial changes proposing the DH126 fitted with twin 3,850lb thrust, rear-mounted de Havilland PS92B jet engines and a 'T' tail based on a fuselage of Dakota dimensions and capacity for thirty passengers. A problem with the 126 project was the lack of a suitable powerplant from DH or any other firm – the PS92B engines were never built and so work on the project slowed through 1961, which was frustrating for the designers.

Hawker Siddeley Aviation

In 1960, at the behest of the Conservative Government, the many British airframe companies merged to form the British Aircraft Corporation (BAC) and Hawker Siddeley Group respectively. At one point de Havilland was about to join the BAC consortium but finally opted for Hawker Siddeley Group (the aviation interests

The HS 136, powered by twin 9,730lb Rolls-Royce Trents with accommodation for fifty-seven passengers, proposed in 1967. There were also longer-fuselage and extended-range versions. (De Havilland Aircraft Heritage Centre model – artwork Rolando Ugolini)

of which became Hawker Siddeley Aviation in 1963). Initially the merger had no tangible effect at Hatfield, which became the de Havilland Division of Hawker Siddeley – but in 1965 the various constituent companies of Hawker Siddeley lost their identities completely. There was also the 'loose end' of de Havilland Engines, which might have been the natural powerplant provider for the new transport, but in 1961, it became part of Bristol Siddeley Engines. Bristol Siddeley was a merger of the Engine Divisions of Bristol, Armstrong Siddeley, Blackburn and DH; and was itself taken over by Rolls-Royce in 1966.

Meanwhile, project work on the DH 126 continued with other engines being considered. In 1964, in order to reduce costs the project was redesigned employing the front fuselage and systems of the Avro (Hawker Siddeley) 748 twin turbo-prop low-wing transport married to a new moderately-swept wing, engines and 'T' tail. This project was designated the HS (for Hawker Siddeley – instead of DH for de Havilland) 131.

But as a thirty seater this was still a small airliner, and the justifiable feeling in the early 1960s was that jet-power was not viable for this size of aircraft. So this project grew into the larger HS 136 with the same basic configuration as the HS 131 but able to carry up to forty passengers. After all the stall problems that the 'T' tailed BAC One-Eleven and Trident suffered from the design was radically re-thought and in 1967 took on another configuration – one that is

Initial Concepts at Hatfield

By 1969 the project had evolved to the HS 144 proposal with a high tail and rear-mounted Rolls-Royce Trent engines. This came in two versions: the HS 144-100 with maximum capacity for sixty-two, and the HS 144-200, capable of carrying eighty passengers. (De Havilland Aircraft Heritage Centre model – artwork Rolando Ugolini)

common to many airliners today. The Boeing 737 and Airbus 320 configuration, twin 9,730lb Rolls-Royce Trents mounted on low wing with a conventional tail and capable of holding fifty-seven passengers.

However, this conventional configuration fell from favour mainly because of concerns over debris ingestion into the engines when operating from unmetalled runways. During 1968 the design grew again with capacity for ninety-three passengers and the project powered by the Rolls-Royce Trent but the project engineers returned to rear-engined, high-tailed design, again known as the HS 136. This design bore a strong resemblance to the BAC 107 design that begat the BAC One-Eleven, a 80ft long, five-abreast seater with two Bristol Siddeley BS75 engines of around 7,000lb thrust, a 500mph cruise speed, a range of 600 miles and a 'T' tail. When the larger One-Eleven was launched in May 1961 it was originally proposed that the 107 should follow one year later but, wisely, the British Aircraft Corporation decided not to develop it, as all the pressure from the One-Eleven's customers was for a larger aircraft rather than a smaller version. The HS 136 was proposed in two versions; HS 136–100 seating sixty-eighty and the larger HS 136-200 with capacity for seventy-ninety.

Meanwhile, at the former Avro Woodford factory near Manchester, which was by then part of Hawker Siddeley's Avro Whitworth division, their design team had proposed the HS 860 as a jet development for the Avro (HS) 748. When the Hawker Siddeley board met they opted for a comparative evaluation of the HS 136 and HS 860. After the evaluation the board told the two teams to work together to develop a final design.

The fruition of the joint group's work was the 1969 HS 144, a revisionist proposal with a high tail and rear-mounted Rolls-Royce Trent engines. This came in two versions: the HS 144-100 with maximum capacity for up to sixty-two and the HS 144-200 capable of carrying eighty passengers. With Rolls-Royce's bankruptcy in 1970, the Trent engine fell by the wayside and once again the project was without an engine. However, there was a powerplant available from a somewhat unlikely source – the American-built Avco Lycoming ALF502 also fitted to the Canadair 600 (the progenitor for the Canadair RJ). Whereas all the previous projects had been based on two engines, the modest 6,700lb thrust of the Avco Lycoming meant that four engines were now needed. As a result the HS 144 configuration developed to a manner similar to the actual 146 with a high-swept wing, four podded engines and a high tail, and a faired double bubble fuselage. Further refinement led to the HS 146 – though it has a circular fuselage section.

Bob Grigg, the 146's Chief Designer, said that he had qualms about whether airlines would buy a four-engined short-haul jet. However, this four-engined layout provided good airfield performance, which meant that certain other costly features such as leading-edge slats and thrust reversers could be omitted on the grounds of weight and complexity. The quietness of the engines is such that no hush kitting was needed and in the years to come this was to be one of the aircraft's major selling points.

Johnnie Johnstone, Director of Marketing, remarked how he loved inviting customers to hear the aircraft take off and how impressed they were by its lack of any roar as the engines were opened up in contrast to all other types. In a similar fashion, the slow landing speeds and quiet braking instead of the deafening reversers makes a 146 landing a pleasant, calm experience.

In researching the market to justify the substantial investment, the Hawker Siddeley Marketing department employed desk research and fieldwork, visiting many airlines and utilising data on all IATA airlines held on a computer at Hatfield. The HS 146 was now identified as a replacement for turbo-prop feeder liners such as the Fokker F-27, Avro 748 and Convairs, providing jet comfort on short to medium sectors without prejudicing field performance. It was to be a simple design that could be managed by unsophisticated operators requiring the minimum of ground equipment. The aircraft could operate from noise-sensitive airfields close to communities and would require minimal ground facilities. The aircraft would have doors at each corner, optional airstairs, an APU (or batteries) for engine starting, and two freight holds with a low sill height.

3

Start-Stop-Start – A Protracted Birth

HS 146 Go-ahead

On 29 August 1973 Hawker Siddeley announced the go-ahead of the project. In order to finance the aircraft the Government invested £46m, to be recovered from a levy on sales. This sum was matched by Hawker Siddeley, which would also bear any cost overruns. However, Sir Arnold Hall, Chairman of Hawker Siddeley, assured the media that cost overruns would not be incurred.

Two basic versions were on offer in two configurations: the 146-100 seating seventy passengers five abreast or eighty-eight at six abreast and the larger 146-200 seating eighty-two or 102 with five or six abreast respectively. The 146-100 was 85ft 10in long; the 146-200 was 93ft 1in long. The wingspan of each was identical at 86ft 6in. As for the engine, there was some surprise at the launch that no British engine manufacturer was involved, especially when the value of the powerplant represented 27 per cent of the machine. It was stated that talks were taking place between Avco and Rolls-Royce with a view to some co-operation at a later stage.

Design, final assembly and flight test of the 146 would be at the Hatfield plant. There would be three development aircraft – the first flying in December 1975 with entry into service in 1977. About 20,000 people in the Hawker Siddeley and supplier plants would be involved in production.

The Hatfield Design team, led by Bob Grigg, designed the aircraft to meet the demanding requirements of the regional air transport market where heavy utilisation over short sector lengths coupled with high reliability are paramount requirements. Outstanding airfield performance and whisper-jet noise levels were other attributes of the aircraft, which offered excellent profit potential on low-density routes. Target price was set at $4.4 million for 1977 delivery, which compared with the twin jet Fokker F-28's price of $4 million.

Michael Heseltine, Conservative Minister for Aerospace, commented at the launch press conference at the Department for Trade & Industry that he was impressed by Hawker Siddeley's commitment to the project and its willingness to invest heavily in it. However, when Sir Arnold Hall, Chairman of Hawker Siddeley, was asked if there was a launching order book, he responded that the firm was yet to seek orders for the 146. He believed that British Airways might have a requirement for a lengthened 146 but that the main market was the global market. In response the British Airways Managing Director, Henry Marking, said that the airline might have a requirement for such an aircraft in the following decade.

At the press conference there was some surprise expressed that this was not an international collaborative project. The response was that Hawker Siddeley was well able to manage the production of a machine of this size without the added challenges of foreign partners. In support of this view, Michael Heseltine declared that it was important for Europe that Britain should maintain a strongly based national aerospace industry.

Fokker Complains

Understandably, there was adverse reaction from Fokker, who obviously saw the 146 as a competitor to their Fokker F-28 and where a senior executive was impertinent enough to claim that the British aircraft was no more than an employment scheme.

In 1962 Fokker launched the Rolls-Royce Spey-powered F-28 as a jet-powered larger companion to its very successful turbo-prop F-27. The F-28 first flew in 1967, entering service two years later. Its configuration was similar to some of the early projects, which developed into the 146, especially the 'T'-tailed, rear-engined HS 136 of the late 1960s. It had a substantial British involvement with its Rolls-Royce engines and Short's manufactured wings. The F-28 was developed into a range of models with initial production versions able to seat fifty-five to sixty-five while later developed versions could accommodate eighty-five passengers. As a type it proved moderately successful, with over 240 produced between 1967 and 1987. (It was further developed into the larger Fokker 100 in the 1980s.)

Financial Times Report

On 30 August the *Financial Times* reported the launch under the headline 'Hawker unveils its good neighbour jet'. The article continued, 'Before launching the HS 146, Hawker spent some years studying the market, with which it already has much experience in selling its twin-engined HS 748 feeder liner turbo-prop, its HS 125 executive jet and its Canadian Twin Otter. It is convinced that the sales potential is there and the initial volume of airline interest (including British Airways Regional Division) tends to confirm the fact. The overall cost of the HS 146 is estimated at £80m including research and development, jigging and

The Hawker Siddeley 146 as depicted in company publicity in 1974, showing operation from a small, unsophisticated airstrip. (Author's collection)

tooling and initial production. Progress payments should be flowing in from about 1975 onwards, if not earlier.'

Towards the end of 1973 work on the 146 gathered momentum, as the firm was aiming for a first flight in December 1975 and certification in February 1977. Initially there would be the production of two batches of six and subsequently batches of ten. The first Series 200 would be the seventh aircraft to fly and this was scheduled for February 1977 with certification in August of that year.

A wooden mock-up was fabricated at Hatfield where design and final assembly work was centred. Other work was split between Hawker Siddeley factories at Hatfield, Brough and Woodford. Woodford was responsible for the wing surfaces, tail and rear fuselage while Brough would build the nose and Hatfield the forward fuselage. No decision was made about the fabrication of the wing though discussions were held with Aerospatiale of France.

Costs Rise

The future for the project was looking good, but within a year events on the world stage, the oil crisis (a product of the Yom Kippur War) and the rampant inflation signalled the worst economic recession since between the wars. Programme costs spiralled to £120 million and as HSA had agreed to bear any overrun it would be

The HS 146 mock-up, built at Hatfield in 1974. (BAE Systems)

liable for substantial sums to the order of an additional £40 million at 1974 prices. It is also highly likely that the return to power of a Labour Government at the February 1974 General Election and their policy to nationalise the British aerospace industry would have had a bearing on any decision made by the HSA Board.

On 17 September 1974 a meeting was held between the management and TASS (Technical, Administrative & Supervisory Staff) trades union at Hatfield where HSA's Managing Director, Mr Thorne, informed them that the board believed that the project was not financially viable. Four weeks later, TASS was told that the 146 would be cancelled and there would be redundancies. union representatives met Tony Benn, the Labour Government's Secretary of State for Industry, and lobbied the Conservative and Liberal Parties for support.

The 146 is Shelved

HSA halted work on 21 October 1974 and informed the Government but agreed with the union to withhold notice of redundancy while discussions continued between the union representatives and the Government. There were work-ins and demonstrations at Hatfield and Brough as workers sought to keep their jobs. Helen Hayman, Labour MP for Hatfield, strongly supported the workers whose factory was in her constituency, and in her maiden speech to the House of Commons spoke in support of the project. The following day, 31 October, there was a mass lobby of the House with 500 staff and shop floor workers, and a delegation visited Downing Street to present a petition. On 4 November, in the House of Commons, Tony Benn said that Hawker Siddeley's decision to cease work was 'a breach of contract' breaking the agreement made between the company and the Government when the 146 was launched in 1973. The action did not stop there, for 2,000 workers from all HSA plants demonstrated at Speakers Corner on 6 November and a further petition delivered to Downing Street. Further support was garnered from various local councils and the then Bishop of St Albans; Dr Runcie (later Archbishop of Canterbury), arranged a meeting for the union delegates with members of the House of Lords.

A Reprieve

On 9 December Tony Benn made a statement in the House that the Government accepted that 50-50 funding was no longer viable, but that they wanted to maintain this civil airliner capability when the aerospace industry was nationalised. Tony Benn said that Sir Arnold Hall had agreed to maintain the necessary jigs, tools, drawings and design capacity.

Tripartite talks took place between HSA, the Government and the union (TASS, later part of MSF and now part of Amicus) under the Chairmanship of Tony Benn and set up a working party to examine the options available. Tony Benn took on an uphill task as the Treasury was also set against continuing with the project. Subsequently, TASS had meetings with the Chancellor of the Exchequer, British Airways and the French Transport Minister to examine if the French wanted to resurrect their interest in manufacturing the wings but were told that France was totally taken up with Airbus.

The workers were trying to convince everybody that without the 146 the state would be taking over a moribund civil aircraft division. This view was not wholly accurate, as there was still the possibility for a further and probably cheaper development of the BAC One-Eleven, which had made over 200 sales but needed a new engine and other improvements to compete with the Boeing 737 and McDonnell Douglas MD-80.

In July 1976 the Government proposed terms to keep the project 'ticking over'. The cost of this was £3.75 million and would finance additional design, research, development and structural testing while the union continued to press for full-scale continuation.

Aerospace Nationalisation and Resuscitation

Lord Beswick, Chairman Designate of the newly nationalised British Aerospace, and other members of the new board arrived at Hatfield on 25 February 1977. He informed the union representatives that on the basis of what HSA had told him, he could not recommend the project to the Minister. In late March, 500 redundancies were announced at Hatfield, but after negotiations these numbers were reduced.

The Nationalisation Bill received the Royal Assent and on 29 April 1977 the state-owned British Aerospace came into being. But it was not the time for the union s to relax their pressure! More meetings were held with MPs, the Labour and Conservative aerospace committees and the British Aerospace Board. So, on 10 July 1978, Gerald Kaufman, as Minister of State at the Department of Industry, announced in the Commons that the Government had given its approval to restart the BAe 146. Victory!

The 'new' BAe 146 would be in two forms – a civil airliner and also a military cargo lifter. The cost was estimated at £250 million which the newly nationalised firm was expected to finance itself.

Sir Raymond Lygo, BAe's Chief Executive from 1986 to 1989, maintains that the decision to go ahead with the 146 was purely political and made by Labour peer and trade unionist Lord Beswick in order to keep the Hatfield plant open. (Hatfield was a Labour constituency while Weybridge was a Conservative constituency.) Perhaps if employees at the former BAC plants had been as organised and single-minded at lobbying their trades unions and Government as those at the former HSA factories, then the BAC One-Eleven might have been developed further, though that would have been at the expense of the BAe 146.

Whatever the reasoning, in the end almost 400 BAe 146/RJs were produced, providing a livelihood for a great number of people in the UK and other aerospace industries from 1978 until the ending of the programme in 2002.

Fokker Complains Again

Just as they had complained at the original launch of the HS 146 in 1973 both the Dutch government and VFW-Fokker complained about the 146's resurrection and the competition it meant for the Dutch-built Fokker F-28. A formal complaint was lodged with the European Economic Community (now the European union) on the grounds that resolutions had prohibited duplication of aircraft types within the Community. Not surprisingly this complaint was unsuccessful.

The Market

At the time of the relaunch, BAe saw the potential market for the aircraft as typically involving low traffic densities over 150-350nm (nautical miles) routes. Airports used would typically be unsophisticated and offer operating difficulties because of their location. They might be situated in remote areas at high altitudes, have challenging obstacle clearance, very short and possibly unprepared runways or, conversely, they might be in city centres where noise would be a critical issue.

After the war, Dakotas serviced many of these sectors, but in the 1950s and 1960s they were superseded by a number of 'Dakota replacements'. None of these wholly dominated the market but the Avro (Hawker Siddeley) 748 Convair 240, 340, 440, and Fokker F27 and NAMC YS-11 generally sold in respectable numbers. On busier sectors in the more developed world, typical airliners used were the BAC One-Eleven, Douglas DC-9, etc.

The market was estimated as more than 1,500 aircraft seating 70-120 passengers (with the bulk in the 70-100 seat range) requiring replacement over the following ten years. BAe's expectation was that it would sell 400 of the type with breakeven at 250 if the American regional airline market were penetrated – which did happen. The overall sales prediction proved realistic, though a large number of these 'sales' were leases rather than outright sales and therefore the aircraft continued to be owned by the manufacturer.

The traffic on many sectors flown by twin-engined aircraft had now grown such that additional capacity was needed with an aircraft of comparable performance and improved comfort providing a competitive advantage for the airline. With the larger airliners – the BAC One-Eleven, Boeing 737 and other twin-jets – British Aerospace's prediction was that noise regulations would drive these aircraft from the skies or require them to have expensive alterations, which would also incur a weight penalty. This analysis did not prove accurate as many airlines fitted 'hush kits' to their aircraft to enable them to meet noise criteria. Even though these cost almost £104,000 in the case of the BAC One-Eleven, most operators carried this out and continued to operate the aircraft.

British Aerospace analysis of this segmented market was that the BAe 146-100, carrying approximately eighty-five passengers with its excellent airfield performance, could replace the twin prop aircraft flying from basic airfields but offering much greater capacity at higher speeds than the existing airliners. The larger 146-200 with 100 passengers could replace the larger types where airfield performance was less critical but still important. Both the variants offered modern airliner comfort with exceptionally low noise levels.

BAe claimed that aircraft/mile costs of the 146-100 would be 20 per cent less than the Boeing 737 and Douglas DC-9 on its optimum stage length of 150nm and where the 146's lower cruising speed would not affect overall times.

The manufacturer expected the average fleet size to be small, from only two aircraft initially, developing to between six and ten. BAe foresaw only a few operators with more than fifteen aircraft. In the early 1980s regional airlines only ordered aircraft in very small numbers because they were unable to raise the money for larger purchases in one go, but would place regular repeat orders.

At the time of the type's first flight, the aircraft's price had more than doubled from the estimate at the original launch in 1974. British Aerospace was asking $10.5 million for an eighty-two-seat 146-100 and $11 million for a 100-seat 146-200. (Both based on six-abreast seating at 33in pitch.)

Initial Sales

The BAe 146 was launched without any orders and as the project progressed there must have been some qualms at the cashflow situation. The firm continued to say that it was the right aircraft at the right time, but it was true that the BAe 146 was not markedly different from the HS 146, in which the Hawker Siddeley board had rightly or wrongly lost confidence.

However, there was much interest in the aircraft but sales were held back by the lack of innovative financing – which became commonplace later. Purchase was a big step for the type of airlines initially looking at the 146, especially with the existence of a strong second-hand market offering aircraft at 10 per cent of the price of a new aircraft. The view was expressed by a former member of the BAe Marketing Department that the aircraft was probably too early for the market in 1981.

In June 1980, LAPA (Lineas Aereas Privadas Argentina) placed an order for two 146-100s and one 146-200 but, as it was unable to obtain the routes planned in Argentina, the orders became options and later lapsed. Four months later, BAe revealed it had two orders from an American airline but did not reveal the airline's identity. Later, this order also lapsed and some other options came and went.

Air Wisconsin Orders the 146-200

On 20 May 1981, the day of the 146's roll-out, a breakthrough was achieved with a launch order from US regional Air Wisconsin for four 146-200s with options on four more. This order was a benchmark in emphasising the significance of the stretched 200 series, which was to prove the best selling version of the type. In choosing the 146, Preston Wilbourne, President of Air Wisconsin, gave the programme a massive and timely boost. Johnnie Johnson, Divisional Sales Director, was quoted in *Flight International*, 'The far-sighted initiative of Air Wisconsin, coupled with US deregulation and the world economy, have totally changed 146 marketing philosophy'. It was now becoming clear how important the larger 146-200 would be – rather than the 146-100.

This order stimulated other American carriers; for example Pacific Southwest Airlines (PSA) followed suit. BAe could then concentrate on the more marketable 200 series rather than the somewhat limited 100 series whose sales were dependent on carriers with special requirements and limited budgets so only likely to order in small numbers.

Instead of the 146-100 as a rather expensive replacement for HS 748s and Fokker F-27s on remote routes, the 146-200 was a way of displacing not only turbo-props but Boeing 737s, McDonnell Douglas DC-9 and BAC One-Elevens. It provided the comfort large airliner customers expected yet it was small enough to operate profitably on lower-density sectors.

Restarting the Project

With the rebirth of the 146, BAe set to work building up the workforce at Hatfield as many had left during the period in which the aircraft was in abeyance. The company also had to renegotiate contracts with suppliers that had originally been agreed in 1973.

In October 1974 Aerospatiale Nantes signed a contract with Hawker Siddeley to produce twelve wing sets but, when the project was restarted, BAe instead negotiated a risk-sharing contract with Avco Aerostructures of the USA. Together with the Avco engines this gave the aircraft a very high American content. Another major foreign contractor was SAAB Scania, which had also agreed a risk-sharing contract to manufacture all the major control surfaces.

The Distribution of Production

British Aerospace plants involved in the production:
- Final assembly – Hatfield 1981-92 and Woodford 1988-2002 (there were twin assembly lines between 1988 and 1992).
- Nose – Hatfield 1981-1993 and Woodford 1993-2002
- Main fuselage – Filton 1981-2002
- Rear fuselage – Woodford 1981-2002
- Fin and flaps – Brough 1981-2002
- Engine pylons – Prestwick 1981-2002

Hatfield built all the noses of the aircraft until closure of the site in 1993. This is a nose test specimen later used for water-tank tests. (BAE Systems)

The first 146 on the production line in August 1980, with assemblies for further aircraft in the background. (BAE Systems)

Other major suppliers:
- Wing – Avco Aerostructures, USA 1981-2002. (In 1992 wing assembly was transferred to BAe Prestwick using parts Avco Aerostructures parts. The RJX wings were assembled at Woodford)
- Powerplant – Avco Lycoming (and successor companies), USA, 1981-2002
- Tailplane, rudder, ailerons, elevators, spoilers – SAAB, Sweden 1981-1992 (Work passed to BAe Prestwick and Chadderton from 1992)
- Engine nacelles – Shorts, Belfast 1981-2002
- Messier-Dowty – Landing Gear 1981-2002

In July 2000, as part of an offset programme, South Africa's Denel Aviation secured contracts from BAE to supply Avro RJ and RJX rudders and ailerons. For the RJX there were some further changes in the supply of components; the engine pylons and nacelles were fabricated by GKN at Cowes.

Production of the 146 gradually built up with deliveries of more than twenty per year between 1986 and 1991 with a peak of thirty-five in 1989. Following two lean years in 1992 and 1993 production stabilised at twenty to twenty-five aircraft per annum from 1993 to 1999. Production then declined steeply to the end of the project.

4
A Technical Description

Aircraft Configuration

The 146's configuration with a high wing and wing-mounted engines has a number of advantages. It provides a continuous clean upper surface, which is clear of ground effect and generates 4 per cent more lift and less drag than a corresponding low wing. Pylon-mounted engines also help to relieve the bending and torsion felt by the wing during flight. In addition the engine pods are well below the wing, reducing drag and keeping the engine efflux clear of the wing trailing edge. This permits the use of highly efficient tabbed Fowler flaps of uninterrupted span, which cover 78 per cent of the wing span and provide a high lift coefficient without the complexity and added weight of leading-edge slats.

The elimination of leading-edge slats reduces the pitch changes that would normally have to be balanced out by the tailplane, so the aircraft has a fixed tailplane – instead of the complexity of an all-moving unit. The 146 has a 'T' tail to provide the greatest moment arm and this allows for the smallest size of tailplane. At the same time the tail position is low enough to avoid any stall problems; it is clear of wing buffet and engine efflux and improves the fin effectiveness by the end-plate effect.

The total configuration gives low approach and touchdown speeds. Touchdown speed at a typical landing weight is only 90 knots and therefore the extra weight and complication of engine thrust-reversers are not required.

A high power-to-weight ratio and good descent capability, combined with short range, means that a modest cruise speed of Mach 0.7 or 310 knots can be used to achieve competitive block times. This leads to the use of a wing with modest sweepback so that simple, low cost, manual controls can be used for the ailerons and elevators.

An advantage of the four-engined layout is in the 'hot and high' performance from short runways. Four engines give a greater reserve of power following an

This view shows the basic 146/RJ configuration – high wing with large one-piece flaps, high tail, large rear air brake, podded engines and a wide undercarriage. The aircraft shown is actually one of the last built – RJX100, G-IRJX (E3378). (BAE Systems)

engine failure – it is far easier for the 146 to clear an obstacle after take off following an engine failure than an equivalent weight twin jet. Other advantages include the ability to make three-engined ferry flights without passengers and the retention of more of the aircraft's electrical and hydraulic power when an engine fails.

Structure

The 146's structural simplicity is demonstrated by the use of conventional materials and mainly high-strength aluminium alloys. Its wing has one-piece skins on top and lower surfaces. Proven assembly methods were employed including thermal bonding of stringers to skins, chemical etching and integral machining. Wide use was made of fully machined structural members, reducing the number of parts and joints.

The entire parallel portion of the fuselage was manufactured in one piece, no joints being necessary where the wing intersection occurs. Also because of the choice of a low cruising speed, no aerodynamic kink point is required in the wing and so the wing was fabricated in one piece with no mid-span joint.

In the event of skin failure there is a failsafe primary structure with crack stoppers and multiple load paths with frames supporting loads. Corrosion protection took place before and during construction. A 180,000 cycle fatigue test programme proved the strength of the structure.

Widespread use was made of standard parts and fasteners. The geometry of the trailing edge of the wing was chosen so that all six lift spoilers are identical and can be interchanged in any position. Similarly, both roll spoilers are identical and can be interchanged from left to the right wing. The flap tracks are non-handed and can be interchanged from side to side. All four engine pods and pylons are identical, as are the main landing gear doors. Identical components bring about a significant saving in the cost of holding spares.

Fuselage Lengths

BAe 146-100 and Avro RJ70 had an 85ft 10in long fuselage.

BAe 146-200 and Avro RJ85 had a fuselage 7ft 11in longer than the smaller aircraft.

BAe 146-300 and Avro RJ100 was 10ft 6in longer than the 200/RJ85.

Because of manufacturing methods the 300/RJ100 was an extended version of the 100/RJ70 fuselage rather than the 200/RJ85. An 8ft 1in fuselage stretch forward of the wing and 7ft 8in aft of the 100RJ70 fuselage created the 300RJ100 fuselage.

Powerplant

The matching of a suitable powerplant to the aircraft configuration was made more critical by environmental noise and pollution level requirements when coupled with providing a competitive fuel economy. The American-built Avco Lycoming ALF502 was selected as the most cost-effective approach. Avco Lycoming of Stratford, Connecticut, was then a division of Avco Corporation. (In 1985 Avco Lycoming was bought by Textron and the company was renamed Textron Lycoming. In 1995 it was sold to Allied Signal, which merged with Honeywell in 1999.)

Although a new civil engine, the ALF502 was based on a design whose engine core had flown over 4.5 million hours in arduous conditions and short -cycle operations in the American Chinook helicopter. This engine was obtained at a very competitive price invalidating the usual argument in favour of twin-engines for short haul aircraft, while the aircraft benefited from the improved performance and the attractive features of a four-engined layout.

The ALF502 is a two-spool turbofan with a high bypass ratio of nearly 6:1 but is a very simple engine with half the number of rotating parts of a typical engine used in twinjet such as a Rolls-Royce Spey or Pratt & Whitney JT8D.

The engine provides an outstandingly quiet operation, still meeting contemporary noise regulations by a substantial margin. The noise regulations for future aircraft are well above the 146 noise output in approach, lateral fly by and take-off, indeed the 146 is even 6db quieter than the current Stage 3 noise minima.

The 146-100/200 was originally fitted with four 6,700lb ALF502R-3. These were upgraded to 6,970 lb ALF502R-3A and later new-build engines were 6,970lb ALF502R-5.

The 146-300 had the 6,970lb ALF502R-5 from the start. Some of the late 146-300s, including all the eight 146-300s delivered to China North West

Airlines in 1992-4 and on G-LUXE (E3001) in its ARA configuration were fitted with the 7,000lb LF507-1H which had hydromechanically controlled fuel control systems like the ALF502.

For the development of the BAe 146 into the Avro RJ in 1992 a development of the ALF502 designated the LF507-1F was introduced. The LF507 had significantly lower operating temperatures for reduced maintenance costs and a higher flat-rated thrust, which enabled improved hot and high operations and higher take off weights. The engine of 7,000lb thrust had an extra low compressor stage and FADEC (Fully Automated Digital Electronic Control), which provided smoother running.

For the final development of the 146 design, the Avro RJX, the advanced Honeywell 7,000lb thrust AS977 was installed and showed substantial potential. The AS977 was promising a growth potential to 9,000lb thrust, on-condition maintenance from the start of service, high reliability with 12,000 hours minimum time on wing and low maintenance costs. Low specific fuel consumption was also predicted though initial fuel consumption figures were greater than had been estimated – possibly if the programme had continued this would have been rectified. Modifications to the pylon were kept to a minimum to allow the AS977 to be retrofitted to existing RJs. Changes include revised engine mounts and minor aerodynamic changes to suit the longer and differently shaped nacelle.

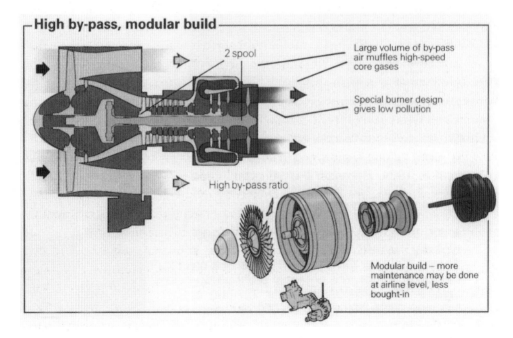

The Avco Lycoming ALF502 powerplant. Its modular build allows for some modules to be changed with the engine still on the pylon. (BAE Systems)

A Technical Description

Auxiliary Power Unit

The APU (Auxiliary Power Unit) is installed in the rear of the fuselage, immediately forward of the airbrakes in a fireproof bay. Access to the unit is via a single panel. The APU provides air for conditioning the cabin and engine-starting and power for systems, and is easily started by a press-button on the flight deck. When on the ground 146/RJs are commonly found with the APU running to provide power – so ground power equipment is unusual.

The AiResearch GTCP36-100 was the original APU on the 146-100 and 200 with the higher capacity -150 fitted to the Series 300. The Sundstrand T-62T-46C-3 APU was introduced as an alternative, initially on the Air Malta RJ70s and has been retrofitted to quite a number of aircraft. However, for the Avro RJX a full FADEC-controlled Hamilton Sundstrand auxiliary power unit APS 1000 was selected.

Flying Controls

The primary roll and pitch controls on the aeroplane are operated by simple low-cost, manual controls, each surface being driven by a simple servo tab. The aileron controls are augmented by a pair of outboard roll spoilers, which are power-operated by single hydraulic jacks programmed from the aileron control circuit. This is to provide a high rate of roll and therefore good manoeuvrability. Three other spoilers on each wing act purely as lift-dumpers to brake the aircraft on landing and are assisted by the outboard spoiler. All the spoilers are hydraulically controlled and can be set to operate automatically on landing.

The only fully powered primary control surface is the rudder, which is the most cost-effective way of obtaining the large rudder authority at low speed with flaps down, but a small authority at high speeds when flaps are up, in order to avoid overstressing the fin.

A feature of both elevator and aileron controls is that they are designed to cope with an inadvertent jam of a control surface or control circuit. In simple terms the captain's control column is connected to the left-hand surfaces and the co-pilot's control column is connected to the right-hand surfaces; both are inter-connected at the flight deck end of the control circuits by spring struts. Thus, for example, should a jam occur anywhere in the left elevator circuit the co-pilot can control the aircraft through the right circuit by breaking out the spring strut.

The rear fuselage-mounted airbrakes are situated on the drag centre of the fuselage. By this means they operate as a speed brake with no change in aircraft trim and no effect on lift as normally occurs on aircraft, which use wing spoilers as airbrakes. The total area of the airbrakes is 40ft^2 and each can be deflected by up to 60°. A feature of this system is that the airbrake hydraulic jack is electrically signalled and this is considerably lighter than a conventional cable-operated type of control.

The lift dumpers shown in the raised position. These operate on touchdown to brake the aircraft. The left aileron and the left outboard roll spoiler are also raised. These operate in unison to bank the aircraft. The flaps are fully extended in this photograph. (BAE Systems)

The Fowler flaps are electrically signalled and can be used in four settings: 18°, 24°, 30° and 33°. A computer monitors the flap movement against pilot selection and in the event of malfunction, i.e. differential deployment, the flaps are stopped.

Flight Deck

The aircraft was launched in 1981 with a conventional flightdeck. It was updated in 1990 when an Electronic Flight Instrument System (EFIS), LED engine instruments together with navigation management and digital weather radar systems were introduced. Beginning with E3163, which first flew in May 1990 all 146s, and the later RJs had these improvements. In 1993 there were further developments in the form of a Category 3 Flight Guidance system and Advanced EFIS.

The flight deck was configured for two-crew operation and much thought was given to reducing pilot workload. The 146 philosophy of elimination of complexity is very apparent on the flight deck. Systems are simple and require little or no in-flight management. Wherever possible the immediate action required following any failure is automatic. The systems instrument and control panels are overhead within easy reach and view of the pilots, with each system grouped on a separate section.

A comprehensive master warning system is installed. In the event of a system failure, a flashing light, either red or amber, dependent upon the degree of

A Technical Description

Another aspect in the aircraft's excellent airfield performance is the large rear airbrake, which can be used in the air and on the ground. (BAE Systems)

importance, will flash in front of each pilot. The display panel on the central instrument panel will indicate the system concerned and the appropriate warning light overhead will give detailed information and enable the appropriate action to be taken.

Flight-deck windows are large to provide excellent visibility, invaluable when operating from airfields in difficult terrain where steeply banked turns or visual circuits may be necessary.

Systems

The guiding principle in the design of the 146 systems was to gain maximum benefit from up-to-date technology without sacrificing simplicity, reliability or low maintenance costs. As an example the four-engined configuration enabled the electrical and hydraulic systems to be designed as mutually supporting systems.

The electrical generators are on the outboard and the hydraulic pumps on the inboard engines. The systems are arranged as such that, with a failure of both electrical generators, a small hydraulically driven emergency power unit obtaining its hydraulic power from the right inner engine automatically supplies back-up power. In addition, the APU installed in the rear fuselage drives a generator identical to those on the engines and can take over in the event of failure of either or both of the main generators.

In a similar manner, the hydraulic system is backed up by the electrical system in the form of an AC hydraulic pump. As an example of design for system

Left: The powerful wing, one-piece flaps and engines on Aspen Airways 146-100 N461AP (E1015). (BAE Systems)

Below: The Electronic Flight Instrument System (EFIS) flight deck, fitted to all aircraft from mid-1990. (BAE Systems)

integrity, the safeguarding of hydraulic power to the brakes is interesting. If the No.3 engine hydraulic system fails, the No.2 engine pump powers them. The AC pump is available if both engine pumps fail and if that has also failed, a small DC pump provides emergency power through a separate system for brakes and emergency landing gear lowering.

Fuel for the 146 is stored in wing tanks and the wing centre section. The wings have an anhedral of 3° but become virtually flat when the aircraft becomes airborne so that fuel feed is not a problem. Fuel is delivered to the engines by electric pumps with back-up hydraulically driven pumps. Extra fuel capacity was available as an option and housed in the rear wing/fuselage fairing. This was fitted to a number of aircraft – notably the three 146-100s operated by the RAF's No.32 (The Royal) Squadron.

In the rear fuselage two conditioning packs, each with a cold air unit, supply the cabin and flight deck distribution systems. They are located to allow easy removal of individual components or a complete pack through an under-fuselage door. Anti-icing for the engines and airframe is by hot air. A hot air bleed from each engine is the main source of conditioning and airframe anti-icing.

Landing Gear

It was envisaged that the 146-100 would operate from unpaved runways and would demand a tough undercarriage, which is exactly what BAe Hatfield designed and Dowty Rotol (later Messier Dowty) supplied. The main legs are a trailing link design with a separate shock absorber to give a soft landing and smooth ride. Despite the fuselage mounting the gear has been given a wide track for stability (which is actually wider than the Lockheed Hercules). Each main leg is enclosed by only one door when retracted and the door is shut mechanically rather than hydraulically. There is a forward-retracting, steerable nose gear with mechanically linked doors.

Cabin Layout

The 11ft 8in outside diameter fuselage cross-section cabin provides plenty of room for passengers and for the carry-on baggage. Customers were offered a wide choice of interior configurations. In early years six-abreast seating arrangements were commonplace in the 146-100 and 200 but from the late 1980s there was a greater emphasis on more spacious five-abreast layouts. The 146-100/RJ70 could carry seventy passengers five-abreast or eighty-two six-abreast, the 146-200/RJ85 could similarly accommodate eighty-five or 100 and the 146-300/RJ100 could carry 100 or 112.

There were plans for a high-density RJ115 capable of accommodating 116-128, which would have required Type 3 mid-fuselage emergency exits to be fitted. Three 146-300s – E3161, E3174 and E3193 – had these installed and later centre fuselage sections were built with structural reinforcements for these Type 3 exits as provision for the RJ115, but these exits were never certified.

Additional or alternative toilet positions, galleys, cabin attendant seats and above-floor baggage stowage were typical options together with different seat pitches and five- or six-abreast seating or a corporate layouts. Between 1983 and 1993, there were over 210 cabin layouts and 140 different galleys variants designed.

The initial overhead drop-down stowage bins fitted to the 146 gave a very neat and spacious look to the cabin but proved flimsy in service. When PSA ordered the aircraft in November 1983 the airline requested bigger bins capable of accommodating golf clubs! These were provided but made the cabin look narrow and encroached on aisle head clearance. These problems were addressed in 1988 with the 'New Interior' with new bins and roof panels taking better advantage of the available space and the fuselage frames were reamed out to improve cabin width by almost two inches. Four years later this interior was superseded on the RJ with the 'Spaceliner' cabin, with further improvements to the cabin sidewalls and an integral handrail introduced just below the bin lid.

The wide fuselage cross-section means that the aircraft in its QT or QC configurations was capable of housing the majority of international freight pallets and containers thus allowing commonality of freight handling with the long-distance cargo aircraft.

The final 'Spaceliner' cabin with wide overhead bins and improved roof panels to give a bright and airy appearance. (BAE Systems)

A Technical Description

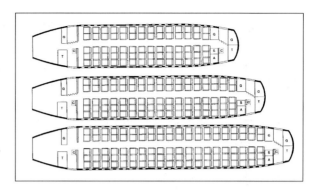

Example seating in a five-abreast layouts for (from the top) 146-100/RJ70 with 70 seats, 146-200/RJ85 with 85 seats and 146-100/RJ100 with 100 seats. These arrangements also indicate Galley (G), Toilet (T) and Crew seat (C). With four doors, there is no need for additional emergency exits. (BAE Systems)

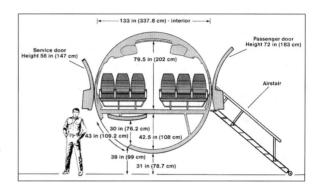

The cabin is the same width as the Boeing 737, and wider than the Fokker 100 and newer Embraer and Bombardier Regional Jets. In a six-abreast layout, the aisle width is typically 16in, with five-abreast generally 20in, but final dimensions are dictated by the seat width. (BAE Systems)

Quick Turn Round

Full-height doors and standard doorsill heights mean that there is no requirement for special air bridge docking procedures. The four cabin doors, two passenger and two service doors, mean the galleys can be serviced while the passengers disembark. The separation of baggage and freight and easy access to the baggage holds also allows rapid handling. The turn round time from engine run-down to the next flight pre-taxi checks is twenty-three minutes minimum compared to a thirty-three minutes minimum turn round time on aircraft with only one passenger door as on the Fokker 100. (Air Wisconsin, a long-established 146 operator, often clocked in twelve-minute turnround times even in the first year of operation.) Another advantage of doors situated at each corner is that there is no need for emergency exits in the cabin with attendant disruption to the seating layout.

The low fuselage enables baggage loading and maintenance to be carried out from ground level. The cargo compartments have centrally located doors thus permitting easy zoning of baggage, thereby reducing turn round times. The forward hold door has been sized such that a fully inflated main wheel or any of the four modules of an ALF502 engine may be carried – a particular attraction to the smaller airline.

A fuselage, stub wing and tail fatigue test specimen being loaded into the water tank at BAe Filton in autumn 1981. In the tank, it was tested to the equivalent of 180,000 flights. (BAE Systems)

Airstairs were installed for most customers at both the forward and rear passenger doors and together with the APU to provide ground power for such necessities as engine starting, cabin air conditioning and lighting, make the aircraft totally independent of ground facilities.

Ground Testing

Hatfield manufactured five structural test specimens to ensure that the aircraft was structurally sound. The structural testing target was 160,000 simulated flights for the four water tank specimens; a fuselage centre-section/wing and a separate nose at Hatfield, and a centre/rear fuselage and a rear fuselage/tail at Filton. About 20,000 'flights' were demonstrated by the time the aircraft went into service in 1983, the equivalent to a year of operation. By 2000 the lead specimen, the fuselage centre-section/wing had completed 180,000 cycles or 'flights'. A further nose specimen was also used for nose landing gear loading and bird strike tests at Hatfield.

An equally comprehensive programme of system tests matched the sequence of tests to prove structural integrity. A full-scale hydraulic and flying control rig was constructed at Hatfield to assess the operating of the systems. Hatfield also produced a complete wooden engineering mock-up to check the fit of all the parts together, plus a complete electrical loom. There were 1,850 hours of wind tunnel resting which utilised twenty models for differing aspects of aerodynamic design.

So BAe were making absolutely sure that when in service nothing would fault the 146!

5

First Flight and Certification

It was a proud moment at Hatfield when the first 146, registered G-SSSH (E1001)1 was rolled out of on 20 May 1981. The roll-out was essentially a PR exercise as the aircraft was far from ready for flight and there then followed a lengthy period of ground testing to check all systems. These involved ground resonance, control stiffness, fuel system and weighing.

The first engine runs took place on 12 August 1981, and as the aircraft was a test vehicle all the test instruments and equipment had to be fitted, checked and calibrated, a lengthy process especially as some pieces were buried in the structure. Staff were impressed by the quietness of the engines and recognised the appropriateness of the G-SSSH registration.

BAe Hatfield-Chester Division Chief Test Pilot, Mike Goodfellow, and his deputy, Peter Sedgwick, spent many hours 'flying the 146' in the simulator. As a result of their many hours on the simulator the pilots had clear expectations of how the 146 would actually perform in the air. They also rehearsed the path of the flight in BAe 125 just prior to the maiden flight.

Maiden Flight

Following high-speed taxi runs and short hops on the evening before, the first flight took place on 3 September at 11:54, when G-SSSH took to the air after a seventeen-second ground roll from Hatfield's runway 24. The crew on the ninety-five minute flight were Mike Goodfellow, Peter Sedgwick, Assistant Flight Test Manager Roger de Mercado and Senior Instrument Engineer Ron Hammond. The 146 was accompanied by BAe 125, G-BFAN which acted as a 'chase plane' during the flight.

The crew ran checks on basic stability and controlability in all three axes, the effects of thrust/speed variations on trim and asymmetric handling with engine number 4 throttled back. The landing gear was raised but the flaps, a

Roll-out of the first 146 G-SSSH (E1001) at Hatfield on 20 May 1981. (BAE Systems)

computer-controlled, single section, were left fixed at 24° throughout the flight. On returning to Hatfield, Mike Goodfellow made two low, almost silent, 300ft passes over the airfield for the benefit of the 4,000 employees and invited guests before landing.

The single-piece Fowler flaps were very large and powerful, but the flap control had proved to be over-sensitive during ground tests and would lock if there were a minute discrepancy during their extension. It was decided to operate them very cautiously and to have them partly extended on the initial flights. Flapless landings were not an option as the landing speed was 50 per cent higher than the flapped. On subsequent flights different but fixed flap positions were employed – only during flight 7 on 17 September were the flaps moved half-speed. The flap system was later modified to provide trouble-free usage.

Hatfield's annual Open Day on 12 September gave the opportunity to show off the latest product to the workforce. By this time, G-SSSH had demonstrated its short take-off and landing (STOL) performance, manoeuvrability and quietness over five flights in a total of six hours fifty minutes in the air.

First Flight and Certification

G-SSSH taking off on its first flight on 3 September 1981. (BAE Systems)

Maiden flight photo call. From left to right: Mike Goodfellow (Chief Test Pilot), Peter Sedgwick (Deputy Chief Test Pilot), Mike Goldsmith, Managing Director, Hatfield, Cyril Braithwaite 146 Project Director, Roger de Mercado (Assistant Flight Test Manager) and Ron Hammond (Senior Instrument Engineer). (BAE Systems)

The first three 146s at Hatfield during the test programme. From left to right, from the front: the second aircraft G-SSHH (E1002) with its right-wing leading edge and engine nacelles marked for icing tests; G-SSSH with a nose probe and a tail parachute, and the third aircraft G-SSCH (E1003). (BAE Systems)

The first period of testing continued until 13 October, by which time the aircraft had made twenty-four flights in fifty hours and G-SSSH was grounded for an inspection. Flight test was justifiably impressed that there was not one delay to departure – the basic design appeared sound.

The Individual Roles of the Test Aircraft

G-SSSH bore the brunt of the test programme required for certification and was joined by the second aircraft G-SSHH (E1002) on 25 January, which took off for a sixty-five minute flight still only in yellow primer. The first few flights were used to compare the aircraft's performance with the prototype, after which it set to work on its own test schedule. The third aircraft G-SSCH joined the others on 2 April 1982. The fourth airframe to fly, G-WISC (E2008) was the first 146-200, and the fifth aircraft G-OBAF, another 146-100, was primarily intended for route-proving.

The main roles of the 146 test aircraft were as follows:

E1001 – G-SSSH – 146-100 (Yellow, orange and brown BAe trim). Handling, stalling, flutter, autopilot, CAA (Civil Aviation Authority) assessment.

First Flight and Certification

E1002 – G-SSHH – 146-100 (Blue BAe trim). Powerplant, systems, icing and tropical trials. These included sojourns in Iceland, Greenland, Torrejon in Spain and Sharjah in the Emirates.

E1003 – G-SSCH – 146-100 (Brown BAe trim). Performance, avionics, noise, crew workload which also required tests in Granada, Spain.

E2008 – G-WISC – 146-200 (Air Wisconsin trim). Testing for 200 series, FAA (Federal Aviation Authority) certification, and noise trials in Casablanca.

E1004 – G-OBAF – 146-100 (British Air Ferries trim). Route proving.

Escape Systems

An escape system was necessary for the high-speed flutter test and low-speed handling tests. For these tests on G-SSSH the four- or five-man crew wore parachutes and helmets and could abandon the aircraft via jettisonable doors. These were the two service doors on the right of the aircraft. The rear door was preferable but as a fallback the forward door was also a possible escape route, though in using this there was an additional hazard that the crew might hit the landing gear if it were extended.

In the event of an emergency escape the crew had set drills, which were regularly practised. To alert the crew there was a system of lights – 'standby – get ready – go'. The co-pilot would dump cabin pressure so that it would equal the outside air pressure. The emergency doors were a normal fitting but without hinges. A crew member would unlock the door and pull a lanyard to lift the door. Suction would do the rest. To assist in escape if the aircraft was manoeuvring violently a rope was strung along the cabin ceiling to assist them in reaching the door. On arriving at the emergency door the parachutes would be clipped to bars above the door and the individual would jump out.

Stall Protection

Though wind tunnel and theoretical tests did not indicate a locked in or deep stall they did indicate that the 146 could pitch up to 40° before recovering. As 'T' tailed aircraft have been prone to serious problems in recovering from stalls and as both a BAC One-Eleven and a Hawker Siddeley Trident and their crews were lost during stall tests, it was rightly decided to fit an anti-stall parachute to G-SSSH.

This was a stall recovery parachute and not a spin recovery parachute, which would have had to be much larger – similar to those carried on the One-Eleven. All that was required for the 146 was a substantial nose down pitch. As parachutes can prove troublesome, a lot of redundancy was engineered into it – duplicated electrical firing and jettisoning circuits and a weak link was incorporated so that the parachute would break free if there were steady drag. The anti-stall parachute

Close-up of the anti-stall parachute installation on G-SSSH, which temporarily replaced the airbrakes during stall tests. The lower part of the fitting contained an explosive bolt which, when ejected, would pull out the parachute mounted above. (BAE Systems)

was fitted in place of the rear air brakes and a nose probe was installed to measure angle of attack and sideslip.

Stalling Performance

The initial intention was to keep the 146 as simple as possible as part of its appeal to the feeder airline market, and so the plan was to achieve a satisfactory stalling performance without the need to install a 'stick pusher' to artificially recover the aircraft.

The flying controls functioned quite simply. The rudder was hydraulically powered, whereas the rest were aerodynamically operated. So BAe spent some time endeavouring to achieve conventional stall performance using triggers on the wing to trip the airflow to make the nose drop without a roll. There was an acceptable stall at most flap angles but at the aft centre of gravity limit the nose down pitch was weak. This could probably have been sorted out with wing fences or vortex generators. However, the engineers realised that this might take too long, delay certification and upset other aspects of the machine's stall performance, so they decided to install a stall identification system with a pneumatic stick pusher: a fairly major modification.

After the loss of a One-Eleven and Trident from 'deep' stalls, the UK requirements were rather more stringent than the American Federal Aviation Agency (FAA). The Civil Aviation Agency (CAA) was rightly determined that there would be no repetition of these fatal accidents. Therefore, the stall protection system had to incorporate a phase advance so that when the aircraft approached the stall at a higher speed, the system would activate sooner and stop the stall beating the system.

General Handling

During the test programme, the general handling of the 146 was assessed as fine but as had been predicted on the simulator the roll control was in need of substantial improvement. Considerable refinement of the lift spoilers operation was needed to achieve design expectations. Roll control is via conventional manual servo-tab ailerons and a roll spoiler (separate from the three lift-dumpers on each wing) operates above the outer lip of the one-piece flaps. Early flying was conducted with ailerons only operating up to 7° of control wheel movement and spoilers were engaged gradually thereafter. The revised functioning used spoilers earlier and less gradually and brought roll control up to required standards. There was no problem with the manual servo-tab elevators or powered rudder, though the pedal force for the latter was eased. Much work was needed on the yaw damper to optimise its performance.

The wing is a powerful lifter assisted by the large fowler flaps but avoids the complication of leading-edge slats and achieves its performance through good design. The runway threshold speed is 100-110 knots and the aircraft's speed is reduced to 80 knots as the nosewheel touches the ground. This low landing speed and the powerful braking provided by the airbrakes and lift dumpers supported the decision not to fit reverse thrust. The undercarriage's smoothness was also universally favourably received.

The 146's airbrakes proved a powerful control – at 150 knots the drag of the aircraft is doubled when they are deployed. However, in contrast, the initial wheel brakes, which were fitted, had carbon pads and these soon started breaking up, so in the short term heavier steel brakes replaced them. In 1984/85, carbon brakes were re-introduced.

Icing Tests

All certification authorities need to be sure that new aircraft can fly safely in icing conditions. Although in theory all icing trials should take place in natural icing, this

G-SSHH and Canberra WV787 in loose formation over Cornwall. WV787 was employed as a tanker to water-spray G-SSHH in icing trials in early 1982. The Canberra had a water tank in the bomb bay connected to a rear spray nozzle. (BAE Systems)

Wet runway trials with G-SSCH in Dan-Air livery at Cranfield. (BAE Systems)

First Flight and Certification

can be a laborious and time-consuming task. For its first operational task between January and April 1982 suitably sensored and equipped (E1002) G-SSHH flew five icing test sorties behind Boscombe Down's water-spraying Canberra WV787.

The Canberra was specially fitted out to carry out de-icing trials on other aircraft, most notably Concorde. It was fitted with a water tank in the bomb bay and a water cloud of up to 90sq.ft at sixty gallons per minute at 200ft range could be sprayed from a long spray-bar that ran under the rear fuselage and from nozzles fitted close under both jet exhausts. A rearward-facing closed circuit television camera was fitted, so that test aircraft flying in the spray could be filmed. The test aircraft also hunted for natural icing around the UK and so in January and February of 1983 G-SSHH was based in Iceland and Greenland for natural icing tests. One of those tests was so 'successful' that when the aircraft landed the crew could not open the doors, which had frozen hard, and had to exit through the forward equipment bay.

First Flight of the 146-200

Sensibly, BAe had recognised the importance of developing the stretched version of the design along with the standard 100 series. However, the first 200 series

The first 146-200 G-WISC (E2008) engaged on slush trials. (BAE Systems)

customer, Air Wisconsin, pushed the manufacturer to produce the prototype G-WISC (E2008) much sooner than originally planned. Originally due to fly in mid-1983, the schedule for G-WISC was expedited and it took to the air piloted by Mike Goodfellow on Sunday, 1 August 1982. The aircraft immediately joined the test programme to examine the effect of the 7ft 10in longer fuselage on handling and performance.

Engine

The engine performed generally well through the test programme – no test flights were held up – drag was well below estimates and fuel burn 6 per cent below expectations. However, a lot of work was needed on the engine thrust management system, which was designed to operate as an autothrottle to reduce the need to make minute adjustments between the thrust levers.

The results of the external noise trials at Granada and Casablanca could not fail to please everybody, for the machine's noise footprint at ninety decibels was one quarter of a typical twin-jet. However, internal noise levels were high and much work was carried out to reduce it to acceptable levels. One area in particular remained untamed and that was 'Flap hoot' – a reedy noise which was audible when flaps were extended from 0° to 18°.

Workload

In December 1982 G-SSCH flew pilot workload trials for the CAA. The aircraft flew a daily 'round robin' from Hatfield, Paris, Amsterdam, London and back to Hatfield. The pilot was rigged up with heart rate monitors and the aircraft had to show that it was simple to operate with different emergency levels. For example, BAe Test Pilot Dan Gurney had to fly a sector while the other pilot simulated incapacity to demonstrate that the machine was easy to fly and systems easy to operate by a single pilot.

Route Proving and Certification

The BAe 146 received its CAA Certification on 4 February 1983 after a comparatively short test programme of 1,500 hours flying compared to other British types such as the One-Eleven, Trident and VC10. It was also the first type certified to the common European Joint Airworthiness Requirements. Four months later, on 20 May, the American FAA awarded a type certificate.

Of the 1,500 hours, 1,300 hours were the usual hard Certificate of Airworthiness testing to worst-case design limits fully occupying the first three aircraft. The remainder were the CAA-approved 175 hours route-proving flown on the fifth aircraft to fly (E1004), crewed by British Air Ferries (BAF) Captains with BAe co-pilots operating out of Southend for a month.

First Flight and Certification

At certification the aircraft's order book was a total of fourteen with options on a further sixteen. Air Wisconsin was taking the lead with the 200 series in the United States while the first British operator would be Dan-Air with the 100 series.

The First UK and International Demonstrations

Farnborough 1982

The BAe 146 was demonstrated in style for its first display at Farnborough International in September 1982. Both the series 100 and 200 were represented with G-SSCH, now in Dan-Air colours, and G-WISC in Air Wisconsin livery in the flying display and G-OBAF in British Air Ferries trim on display statically.

Indian and Far East Tour

Even before the 146 had been certified, it was decided to send 146-100, G-ASCHH (E1005) on a sales tour to India and the Far East. Finished in the same livery as the prototype, it made its first forty-minute flight from Hatfield captained by Peter Sedgwick on Saturday 19 October 1982. On the following day, it made two more flights to clear any snags. Flight Test did not want to clear it and very nearly delayed its departure, but it left on time. For the tour, G-ASCHH was fitted with seventy-four seats and a large galley in order to provide a high level of comfort for any potential buyers.

G-SSCH demonstrating the 146's airfield performance at the aircraft's first Farnborough Air Show in September 1982. (BAE Systems)

G-SCHH (E1005) overflying Sydney during the Indian and Far East Sales Tour of October –December 1982. (BAE Systems)

Meanwhile, Peter Sedgwick and the tour crew flew by British Airways to Dubai while others at Hatfield checked out the aircraft and duly delivered it to Dubai. The 146 then staged via India, Thailand and Hong Kong to Japan where it spent eight days on intensive demonstrations to Southwest Airlines, TOA Domestic Airlines and All Nippon Airways. Japan had been identified as a strong potential market with a highly developed system of domestic air services using many small airports. (Six years earlier, BAC had demonstrated a specially developed short-field One-Eleven to Japanese airlines.) The 146 flew into Osaka on 5 November to carry out noise measurement trials. The Japanese set up twenty-three noise monitoring devices and measured the 146 against the other aircraft present.

The 146 proved all BAe's claims of being the world's quietest airliner, profoundly beating all the other aircraft present (Douglas DC-9, Boeing 727 and 737, Airbus A300 and Lockheed Tristar) by a very substantial margin. In fact, the results for the British aircraft operating with a typical payload was 76.6 decibels, whereas most of the other aircraft's results were in the 90 decibel range. The 146 with its superior STOL performance seemed a logical replacement for the Japanese-built twin Rolls-Royce Dart turbo-prop NAMC YS-11, which was twice as loud as the British jet airliner.

First Flight and Certification

The 146's stay in Japan coincided with Prince Philip's visit in his capacity as President of the World Wildlife Fund and the Royal party used the aircraft for two days. Prince Philip had the opportunity to fly it, though not to land it as the UK CAA had stipulated that as the 146 was not at that time certified, there always had to be a BAe pilot in command.

After the demonstrations in Japan, the 146 set off for the Philippines, where it was demonstrated to executives of Philippines Airlines in Manila. The 146 was only the second jet to fly into the Philippines' summer capital Baguio – a critical airport with a 5,512ft runway on a high plateau.

From the Philippines, the demonstrator headed south for Perth, in Western Australia, arriving on 10 November for a week-long series of demonstrations on typical routes to the airlines such as TAA and Ansett and the Government. From Melbourne, G-SCHH then flew the longest flight of the tour – 1,500 miles to Christchurch, New Zealand. For three days, from 17 November, the registration ZK-SHH was assigned (though never carried) owing to local regulations with regard to aircraft as yet without a Certificate of Airworthiness. The five-day stay was a continuation of intensive demonstrations, including landing at Mount Cook airfield over 2,000ft above sea level in gusts of up to 40mph and which could only be approached through the mountains.

From Auckland, Peter Sedgwick and his crew piloted the 146 to Port Moresby in Papua New Guinea where it was demonstrated to Air Niugini. This was the easternmost part of the tour, which was, however, far from over. Arriving in Kuala Lumpur on 25 November, it was demonstrated to Malaysian Airlines and then on to Delhi where it flew into Safdarjang, Delhi's city-centre airport with a one-way 3,000ft runway. G-SCHH made various flights for Indian Airways Corporation and Vayudoot before continuing on its route home via Pakistan, Muscat and Kuwait.

The demonstrator arrived back at Hatfield on the late afternoon of 13 December after flying 58,800 miles over fifty-one days and carrying 3,300 passengers. This proved a very successful tour, as the aircraft, still uncertified, was very immature (for instance, it did not have a stall warning system). Though there was an APU failure at Hong Kong and the flap computer twice delayed departures, the serviceability and remarkable quietness impressed all.

The Japanese, though very impressed, did not order the BAe 146, and the type has never served on the Japanese register. But orders were to follow later from Australian, New Zealand, Thai and New Guinea airlines.

After all the hard work that had gone into the preparation and organisation of the tour a great fillip was the announcement after the aircraft's return that the RAF was to purchase two 146s in the following year for an evaluation as to their suitability for the Queen's Flight.

From December 1982 until January 1983, G-OBAF (E1004) was engaged on 175 hours of route-proving trials with British Air Ferries, which was necessary for certification. Following this, it departed on an eighteen-day tour of Africa, beginning in Egypt. (BAE Systems)

African Tour

Following the route-proving exercise, G-OBAF was immediately readied for a tour of Africa. There was just time to make a subtle alteration to the BAF titling on the fin and change it to read BAe. The aircraft departed Hatfield on 6 February 1983 for Egypt where it spent two days before flying south to Harare in Zimbabwe. Operating from the Zimbabwean capital, there were demonstrations in Maputo in Angola, Luanda, Swaziland, Zambia, Malawi and Johannesburg. The landings at Kariba in Zimbabwe were always interesting, as ATC warned of elephants on the runway! Dan Gurney, piloting the aircraft, said, '… the Maputo demonstration flight could not be flown overland owing to risk of ground fire, and similar cautions prevailing in Luanda, but we were carrying the new UK ambassador, who confidently said that this provided some protection.' G-OBAF returned to Hatfield on 24 February via Nairobi, Khartoum, Cairo and Brindisi to complete its 24,000-mile tour.

6

The Stretched 146-300 Series

British Aerospace showed great perspicacity with the 146 by launching the 100 and 200 series at the same time and ensuring that their flight test programmes were completed within three months of each other. There must have been some doubters as the very development of the larger, heavier 200 was a move radically away from the originally envisaged small feederliner. The expectation at the time of launch was that the 200 would take just 30 per cent of all sales. Yet when production ended in 2002 the final breakdown of series production was as follows:

- 47 x 146-100/RJ70 = 12 per cent
- 203 x 146-200/RJ85 = 52 per cent
- 142 x 146-300/RJ100 = 36 per cent

So the decision to produce the 200 series early on was obviously the right one.

Even before the first flight of the prototype in September 1981, Hatfield Future Projects had been working on various developments of the 146 including a lengthened 146-300. By 1984 the 146 marketing effort began to address itself more to big-league airlines.

The launch of the Rolls-Royce Tay-powered Fokker 100 in November 1983 was seen as a potential threat to the 146 and Swissair's order for it a few months later seemed proof of that. The Fokker 100 was a stretched and thoroughly modernised version of the twin-engined 'T' tail F-28 Fellowship jet airliner with seating for a maximum of 122 and a modern EFIS glass flightdeck. It first flew in November 1986 and was certified a year later with the first customer delivery to Swissair in February 1988.

BAe announced provisional details of the 146-300 at Farnborough 1984 and Divisional Sales Director Johnnie Johnson said that it would be eminently practical to stretch the 146 while retaining the same performance as the 200 series.

BAe's challenge was to raise launch aid from the Government, which only earlier in the year had had to have its arm twisted before granting BAe £250 million aid for the Airbus A320 (BAe had to find £200 million itself). Money was also being spent on other projects – the ATP (BAe 748 replacement) and on the EAP fighter demonstrator.

An early projected 300 series devised in 1984 had a fuselage stretch of 10ft 11in greater than the 100 series, more powerful Avco Lycoming ALF 502R-7 engines at 7,500lbs thrust, winglets and a 'glass' cockpit. Two years later, there had been considerable refinement and some simplification in the design. The final stretch was greater than initially proposed at 15ft 9in, the standard engine of the 100 and 200 series was used and the winglets and 'glass' cockpit were deleted.

Fortuitously, the designers of the bantamweight 100 series gave it a wing of almost Herculean capacity for operation from short, hot and high airfields. Not only did the same wing lift the heavier 200 series with ease, but it could also accommodate the weight of the even heavier 146-300.

In 1984, BAe referred to the 300 series as a six-abreast 120-seater, yet by the time the prototype took to the air it was promoted in a 100-seat, 32in-pitch, five-abreast layout with a 20in aisle. British Aerospace had refocused on the needs of the discriminating passenger, who had become a feature of the maturing regional airline market – particularly in the USA, where the five-abreast layout on the type was already popular.

Likewise the 200, originally devised as a 100 seater with a six-abreast layout was marketed with 85 seats in a five-abreast arrangement. So the new 300 could accommodate the passenger load of 200 more comfortably with the similar airfield performance and quietness.

Features of the 146-300
The differences between the earlier marques was as follows:
- The 146-300 fuselage was a 100 series fuselage with an 8ft 1in plug forward and 7ft 8in plug aft of the wing
- This stretch increased passenger capacity to 100 at five abreast and 112 at six abreast. Capacity could have been raised to 128 if type 3 escape exits had been added at the middle of the cabin. (Three 146-300s were fitted with these exits but these were never certificated.)
- Thicker fuselage skin strengthened the 300 at the mid-section.
- The reshaping of the 300's frames and side panels added nearly two inches to internal width.
- 'New Interior' cabin styling featured larger luggage lockers, with doors tailored to give better access and allow light inside.

The Stretched 146-300 Series

- Interior noise reduction was the result of improvements to the wing fillets, door seals and the silencing of a hydraulic pump.
- Strengthened top wing skin and stringers of new material.
- The landing gear was modified, and carbon brakes were made standard.
- The powerplant remained the Textron-Lycoming ALF502R-5 while the engine pylon attachments underwent minor changes.

By this time the engine was now nearing two million hours and cycles on the 146. Since the initial learning curve, in-flight shutdowns had dropped to about 0.01 per thousand hours. Unscheduled removals were also down to 0.2 per thousand hours, but since engine removal was easy, engineers often preferred to take the engine off rather than take advantage of the modular construction.

The New Assembly Hall

A BAe executive who was very influential in driving the 300 forward was Charles Masefield, who became Divisional Director and General Manager responsible for Hatfield in April 1986. He campaigned for 300 and won new investment for the Hatfield site with its new assembly hall, which was agreed in June 1986 and cost £4 million to build. It would improve efficiency, as the aircraft with wing, fin, tailplane and undercarriage attached would be towed

Stretching in progress at Hatfield. The 100 series prototype, G-SSSH, is shown cut into three and awaiting the insertion of two fuselage sections, one forward and one aft, to become the 300 series prototype, with the new registration of G-LUXE. (BAE Systems)

to the new assembly hall where they would be backed into one of four fixed docks and completed, whereupon they would move to flight test. This would eliminate the time wasted in the typical regular production line moving forward and requiring the laborious manhandling of staging.

As the roof supports in the existing assembly hangars at Hatfield were low, once the tail was attached, moving the aircraft required a delicate operation with the nose jacked up and the fuselage at an angle. This investment in the new assembly hall would pay for itself, as it would reduce assembly time by four weeks, besides offering better working conditions and improving the image of the factory and firm.

Converting G-SSSH into the 146-300 Prototype

G-SSSH (E1001) made its last flight as a 100 on 7 August 1986 having flown 1,239 hours. Approval to start work on the conversion of E1001 to the aerodynamic prototype of the 300 series was given in late August 1986. The need to have the aircraft flying within eight months to enable it to attend the Paris Air Show clearly demanded exceptional measures. Management saw that the only way to achieve such a timescale was to reintroduce the concept of a self-contained team working in a 'one off' shop. The 'experimental' bay of the Technical Services Department was chosen, where the prototypes of the DH Hornet, 108, Comet, 110, Dove and Heron had all been built. An essential

Prior to being repainted, the 146-300 prototype clearly shows its front fuselage plug still in primer. (BAE Systems)

The Stretched 146-300 Series

First flight – 1 May 1987. As an inspiration to the workforce, Charles Masefield, Hatfield General Manager, had the date and time of the 146-300's maiden flight boldly painted on the doors of Technical Services. (BAE Systems)

factor in the smooth progress of the conversion was the positioning of design and engineering teams, complete with drawing boards and terminals, adjacent to their production colleagues.

The 146-300 team under Project Manager Tony Fairbrother certainly lived up to the very highest traditions of their famous surroundings and proved that the firm could match the spectacular timescales achieved in years gone by, despite more demanding requirements. The aircraft was cut into three and two fuselage plugs inserted to lengthen the aircraft. It was rolled out with the new sections still in primer on 8 March 1987 but for its first flight was repainted and re-registered as G-LUXE. (Its construction number now became E3001 as its fuselage length had increased from a 100 to a 300 series.)

First Flight – 1 May 1987 and the 'Sounds of Silence'

Charles Masefield decreed that the 146-300 would fly May Day at 12:00 and to engrain this idea in the mind of the workforce he had *'First Flight – 1 May 1987 – 12 Noon'* painted on the door of Technical Services. On May Day, the new Assembly Hall was formally opened, and then, inside the hall, a curtain drew back to reveal G-LUXE while a choir sang Simon and Garfunkel's the 'Sound of Silence'. The 146-300 was then rolled out and prepared for its 'second' maiden flight.

G-LUXE, taxiing back to the Hatfield apron after its successful maiden flight on 1 May 1987. (Derek Ferguson)

Maiden Flight of the 146-300

Punctually at noon in front of 5,000 guests and employees, G-LUXE took off on what proved to be a snag-free first flight. On landing, the crew, Chief Test Pilot Peter Sedgwick, Deputy Chief Test Pilot Peter Tait, Flight Development Engineer Dave Gibbons and Instrumentation Engineer Ron Hammond, were greeted by Sir Raymond Lygo on the aircraft's steps.

On that day, the Civil Aircraft Division received the Queen's Award for Industry for the second year running and an order was announced from Air Wisconsin as launch customer, for five 300s to join the ten 200s it already had in service. A few weeks later, BAe promoted the 146 in force at the Paris Air Show with G-LUXE in the flying display, a Continental Jet Express 200 in the static display and another 200 flying potential customers in and out.

By late August, G-LUXE had flown seventy hours of testing in forty-six flights which included general handling at an aft centre of gravity, autopilot and yaw damper calibration, flutter tests, minimum speed take off tests and measured take offs and landings. Meanwhile, on the production lines, the first production 300 series fuselages where taking shape. Testing continued apace, and by March 1988 some 250 hours had been flown including some high-speed and high-weight flights. G-LUXE was then prepared for stall handling so a stall parachute was fitted in place of the rear air brakes. By the end of April, these tests were satisfactorily completed so the parachute was removed. Tests continued with more airfield performance trials, including accelerate/stop landing distance measurements.

Pilots found there was virtually no difference in the handling of the 100, 200 or 300 series, except that the 100 was slightly sporty and the 300 was more sedate.

The Stretched 146-300 Series

Sir Raymond Lygo, British Aerospace Chief Executive, on the aircraft's steps, greeting the crew of the 146-300 after its maiden flight with Peter Sedgwick, Chief Test Pilot, on his right, and other members of the flight crew and dignitaries behind. (Derek Ferguson)

The partially completed first production 146-300 (E3118) outside the new Hatfield Assembly Hall. This aircraft flew on 22 June 1988 registered as G-OAJF, and was used for test purposes until delivery to Crossair. (BAE Systems)

First Production 146-300

All employees were invited to watch the maiden flight of the first production 146-300 G-OAJF piloted by Deputy Chief Test Pilot Peter Tait with Alan Foster as co-pilot on 22 June 1988, two days ahead of schedule. The aircraft was airborne for two hours and fifteen minutes. It was registered as G-OAJF as a mark of respect to Tony Fairbrother (Anthony James Fairbrother) who had led the conversion of G-LUXE. Unlike G-LUXE, which was produced by lengthening the fuselage of the 100 prototype and consequently differed markedly to then current production standards, G-OAJF was built to production standards and could perform at 300 series operational weights.

A further flight was made the next day before it was hangared for flutter trial installations. Together with G-LUXE, the 300 series had now clocked 350 hours. (Along with G-OAJF's maiden flight in June 1988, three other 146s took to the air for the first time – the first time Hatfield had achieved four 146 first flights in one month.)

146-300 Certification

Certification of the 300 series was announced on 6 September 1988 at the Farnborough Air Show. G-LUXE took part in the flying, and the second production 300 (E3120) which had flown at the end of August was in the static display in the colours of United Express but with the British registration G-BOWW. Also on static display was G-BSTA (E1002) in its latest guise as the Small Tactical Airlifter (STA) promoting the four military versions of the basic design announced at the 1987 Paris Air Show.

G-OAJF flew with Electronic Flight Instruments Systems (EFIS) on 20 March 1989 and was soon certified. EFIS became standard in the following year on all 146s with airframe E3163, a Hatfield-built 146-300. G-OAJF remained with BAe for flight testing until early 1991, when it was refurbished at Hatfield and was leased to Crossair HB-IXZ in September of that year.

If British Aerospace had any doubts about the wisdom to stretch the aircraft, the announcement of the sale of nineteen BAe 146s with the greater majority for the new stretched 300, should have confounded any criticism.

Two 146-300s in final assembly in the Hatfield Assembly Hall, which was opened in 1987. (BAE Systems)

7

The 146 in Service

First Services with Dan-Air

British Aerospace had fought to achieve a sale to Dan-Air and the contract was signed on 23 September 1982 for two 146-100s with options on two more. On 23 May, G-BKMN (E1006), fitted out in an eighty-eight seat layout, was ceremonially handed over to Mr Newman, the Dan-Air Chairman at Hatfield. Four days later, it operated the first commercial service of the 146 from Gatwick to Dublin.

From that day, G-BKMN averaged more than eight and a half hours of flying day in, day out. It flew 600 hours in the first sixty-nine days, starting out most days at 07:00 to fly to Dublin. The return flight was normally followed by the 11:45 departure to Berne, and on return from that round trip the aircraft worked the 16:30 Toulouse flight. The return from there was scheduled for 20:30, after which the 146 often flew tourists to Venice or Gerona. The low noise signature of the aircraft meant that Dan-Air could plan a return at 03:00 without disturbing the locality.

At the time, the BAe 146 was the only jet in the world certified to operate from Berne's 4,000ft runway. However, Dan-Air's second 146 G-BKHT had the misfortune to run off the runway at Berne after an aborted take off on 23 November 1983, but soon returned to service. Initially, the duo was based separately at Gatwick and Newcastle respectively, serving Berne, Bergen, Dublin, Stavangar and Toulouse on scheduled flights. After six months' service, the new British airliner had performed well – almost 3,000 hours had been flown and 2,220 landings made, with an average utilisation of 6.6 flights per day. In the first month, 12 per cent of flights were delayed but by November that figure had substantially improved and was down to 2.5 per cent. The 146 also demonstrated its economy by burning 16 per cent less fuel than the BAC One-Eleven 200.

Additional inclusive tour destinations that soon profited from the 146's performance were Innsbruck and Chambery, neither of which had been served

The first 146 delivery was G-BKMN (E1006) to Dan-Air on 23 May 1983. It flew the first passenger service on 27 May 1983. After service with a number of operators, it now flies as a corporate aircraft for Formula One as G-OFOA. (BAE Systems)

by jets previously. In July 1984 the Dan-Air duo was joined by G-SCHH and, six months later in January 1985, Dan-Air celebrated its one millionth 146 passenger.

From 1989, Dan-Air began to trade in its 100 series for the longer 300 series. Five of these were delivered between 1989 and 1992, but with Dan-Air's demise in November 1992, all the aircraft were returned to British Aerospace and stored at Filton or Hatfield before redeployment.

Air UK to Buzz

Air UK was a merger of several British carriers and by 1988 had moved its headquarters to Stansted, by which time it was the third largest British airline with twenty turbo-props. Simultaneously, it began the gradual re-equipment of its fleet by initially ordering two 146-200s in November 1987 and receiving the first almost immediately when 146-200 G-CNMF was delivered on 27 November 1987. It was soon joined by two more new 200s and two second-hand examples.

The 146 in Service

When Ryanair took over Buzz in 2003, it kept four 146-300s until the end of the year. This is G-UKRC (E3158) operating a Ryanair service from London Stansted to Blackpool, but still entirely in Buzz livery. Note the airstairs fitted at the front and rear doors, which fold upwards and stow inside the aircraft. (Author)

The first production 146-300 destined for a European carrier was delivered to Air UK as G-UKHP on 28 February 1989, soon followed by the second G-UKAC. They were introduced on Air UK's new routes between Gatwick and Scotland on 2 March. Two early 146-100s also joined the fleet in 1989. So in the early 1990s it was the only UK airline employing all three versions of the 146.

Air UK gradually increased its fleet to ten 300s and was then taken over by KLM to become KLM UK in 1998. This metamorphosed into a budget airline called Buzz in 2000 with eight 146-300s, all painted yellow. Buzz did not survive in the predatory low-cost market for long and was taken over by Ryanair in 2003, which kept four of the 146s for the summer schedules and then disposed of them by the end of the year.

Debonair

As one of the pioneers of the new European low price airlines, Debonair operated scheduled low-cost services out of Luton to European destinations. It operated a total of eleven 200s and two 100s between 1996 and 1999 when the airline went out of business.

Originally delivered to PSA as N351PS (E2028) in 1984, this 146 was withdrawn from service and stored in the Mojave Desert for more than four years before refurbishment at Marshall Aerospace, Cambridge, in 1996. It then flew with Debonair as G-DEBA from 1996-99. (Marshall Aerospace)

Their first aircraft, G-DEBA, delivered in May 1996, was formerly N351PS of PSA and N171US of USAir and had spent five years parked in the Mojave Desert. Most of its other 146s were also former PSA machines, which had previously been parked in the Mojave.

Flightline

Southend-based Flightline operates no services of its own; instead it provides airline and corporate wet-leasing services. Flightline BAe 146s have operated services for British Airways, Swissair, Aer Lingus, Croatia Airlines, Lufthansa, Air France and others. The carrier is also regularly engaged for corporate work for the Oil Industry, Premiership Football Clubs, car manufacturers (e.g. Jaguar) and the Conservative Party. It first started using the 146 in March 1993 with the arrival of G-BPNT and by 2004 the 146 fleet had increased to five 300s and six 200s.

Flybe

Flybe, formerly named British European and before that Jersey European, has become Britain's third largest domestic carrier. In 2002 the airline rebranded itself as 'Flybe' with a new blue livery and adopted a market segment between the full fare and the low cost airlines. Its major hubs are Southampton and

Exeter, but it also operates from Luton, London City and Gatwick and its biggest route is Gatwick to Belfast City airport. A significant development in 1996 was the franchise deal with Air France where British European 146s in Air France colours connected Birmingham, Edinburgh and Glasgow with Paris.

British European originally introduced 146s in April 1993 and currently operates fifteen of them. The airline has a technically mixed fleet of 100s, 200s and 300s. Five of its six 300s (former Thai Airways machines) have EFIS flightdecks, while the other does not, and its 200s have two different types of undercarriage.

As part of its expansion plans, British European decided to expand its 146/RJ fleet and on 1 March 2001 ordered twelve of the RJX100 development of the type with options on a further eight. First deliveries were to be in April 2002 stretching on until 2006. However, with the sudden cancellation of the RJX in November 2001 the carrier had to re-equip with other aircraft and bought Bombardier turbo-prop Q400s.

CityJet and Aer Lingus

Some 146s have operated with a number of operators and worn many liveries, and this is very much the case for those operated by the two Irish airlines – CityJet and Aer Lingus. Carrying one million passengers per year, CityJet began operations in 1994 with six franchise routes on behalf of Air France between Paris and Dublin, London City, Edinburgh, Gothenburg, Zurich and Florence. CityJet is a 100% owned subsidiary of Air France and operates twelve BAe 146-200 series aircraft, all of which at one time served in the United States with either PSA, Presidential, Air Cal or Westair.

Between 1995 and 2003 Aer Lingus was also a 146-300 user on routes from Dublin to UK regional airports. In a move to cut costs, it decided to reduce the number of types in the fleet and withdrew them from service. Aer Lingus's eight aircraft had previously been with Thai Airways, Sagittair, Dan-Air, British World, and Air UK. They are now mainly inactive and three of them, EI-CLG, -CLH, and -CLI are in store in the Mojave Desert.

German 146s

In 1998, WDL bought a 146-100 from BAe Asset Management for ad hoc and passenger charter work. These services are operated both on behalf of other airlines as well as in a Corporate Shuttle role. Airline customers have included Air Berlin, Air Nostrum, Lineas Aereas, Eurowings, Aero Lloyd, Hapag Lloyd, Virgin Express, FTI, and LTU. Corporate flights are carried out for organisations like BMW, Audi, Daimler Chrysler and Airbus Industries. To cater for these

First flown in 1983, German Airlines WDL's appropriately registered D-AWDL (E1011) previously operated with four other airlines: TABA, Royal West, Air Nova and Air UK. (Author's collection)

operations the fleet has gradually grown to four aircraft and the original 100 has been joined by two 200s and a sole 300 series.

In contrast, Eurowings, created from a merger of two German regional airlines, serves fourteen scheduled destinations in Germany and twelve in Europe. It has a fleet of ATR turbo-props, Airbus 319s, four 146-200s and four of the larger 146-300s. The 146s and Airbuses share the high density and longer sectors for the airline. In 2003 it became a member of Lufthansa Regional and expanded its 146 fleet.

Atlantic Airways

In November 1987, Atlantic Airways was established to run services to and from the Faroe Islands. After considerable examination of the market, the management chose the BAe 146-200. With the delivery of OY-CRG formerly with PSA, Faroese people were able on 28 March 1988 to fly for the first time from Vágar (the third largest of the Faeroe Islands) to Copenhagen. In June 2000 a second 146-200 OY-RCA, formerly with PSA and Debonair, joined Atlantic Airways to assist in serving the burgeoning schedule that now includes Billund in Denmark, Reykjavik, Aberdeen, London Stansted and Oslo. Atlantic Airways have added two 146-200s to their fleet during the last two years.

Into Service in the United States

British Aerospace was fortunate in obtaining a launch customer for the 146-200 and the 146-300 series of the calibre of Air Wisconsin – which continues to operate the type to this day. On the West Coast of the USA, the aircraft also quickly attracted large orders from PSA and Air Cal, but when these airlines were taken over by larger carriers, the 146s fared less well and were withdrawn from service. Deregulation brought about massive changes in the structure of the American regional carriers which now fed their services into dominant franchise partners at their hubs and in return gained the outbound traffic. For example, in the late 1980s, the United Express Group incorporated several 146 operators – Air Wisconsin, Westair, Aspen and Presidential.

Air Wisconsin

The BAe 146's launch customer, Air Wisconsin, was formed in 1965 to provide services between the airline's base in Appleton and Chicago, Illinois. Air Wisconsin was the first US regional airline to introduce regional jet services with a fleet of British Aerospace 146s and developed a route network that stretched from Colorado to the USA's eastern seaboard. Though its main base was at Appleton, it established a new maintenance base for the 146 at Fort Wayne, Indiana. The airline influenced British Aerospace to certify the 200 series sooner than would otherwise have been the case and also added much of their expertise, for example to the cabin seating design, which was then adopted for the aircraft in general.

Air Wisconsin has the distinction of being the 146-200 and 146-300 launch customer. As a franchise partner of United, its aircraft operate in United Express livery. Here is 146-300 N612AW (E3122) at Hatfield, which was delivered in December 1988. (Derek Ferguson)

First US Delivery

The first production aircraft for Air Wisconsin's initial four aircraft, N601AW (E2012), made its maiden flight on 25 May 1983 from Hatfield to Woodford for passenger evacuation trials. It left on delivery on 16 June routing via Prestwick, Keflavik – Sondestromfiord (Greenland) – Goose Bay – Green Bay to Outagamie Airport Appleton, Wisconsin. The newly delivered aircraft flew its first revenue service on 27 June with a thirty-two-minute service from Fort Wayne to Chicago. Very much as Hatfield's market research had predicted, the 100 seater 146-200s replaced propeller-driven de Havilland Canada Dash 7s – offering double the capacity and faster sector times on average stage lengths of 127 miles.

In Service

From the start, the N601AW was heavily utilised, flying an arduous thirteen sector daily timetable. The wide cabin 146 was a major improvement over the Dash 7s in passenger appeal and the seat mile costs were three times better. Break-even load factor was 36 per cent so the airline was happy, as the average load factor was 47 per cent. Average sector time was about thirty-one minutes and each airliner repeated the task twelve times per day with turn rounds of often only twelve minutes. 'Achieving 4,200 cycles per annum per aircraft with a regional airliner is unbelievable', said Johnnie Johnson.

With four aircraft in operation by February 1984, Air Wisconsin soon 'firmed up' its options and a further four aircraft were delivered between 1984 and 1986. The ninth and tenth 146-200s were both delivered in the following year. Air Wisconsin developed into a full code-sharing arrangement with United Airlines in 1985, and its aircraft were repainted in United Express livery.

At Hatfield on 1 May 1987, the day of the 146-300's first flight, an order was announced by Air Wisconsin for five 300s with an option on ten more. By June 1988 the first Air Wisconsin machine had completed the highest number of flights – over 16,600 with an average duration of thirty-one minutes.

Air Wisconsin Introduces the 146-300

Air Wisconsin took delivery of their first 146-300 N611AW (E3120) painted in the livery of United Express on 16 December 1988 after it had been used as a BAE demonstrator in the USA with the special registration N146UK. The second aircraft of the order, which was delivered before Christmas, closely followed. The remainder of the five on order were received by Air Wisconsin during the next year, giving them a total fleet of 15 BAe 146s plus Aspen's three 100 series. Air Wisconsin was wholly owned by United for a short period, but in 1993 regained its independence, though it continued its connection with

United. The airline currently has one series 146-100, twelve series 200s, and five series 300s.

Air Wisconsin and Scope Clauses

A limiting factor in the US market is the pilots' union scope clauses, which effectively determine the seat capacity for aircraft operated by regional airlines and those flown by the national carriers. So the scope clauses have limited the regional market. Aircraft larger than seventy seats are flown by the major airlines with unionised crews, but non-unionised and therefore less well-paid crew can operate aircraft with fewer seats. In this constrained environment, the BAe 146 was one of the more successful Regional Aircraft in the 1980s and 1990s with Air Wisconsin and its eighteen-strong 146 fleet. Air Wisconsin negotiated a deal whereby it may operate eighteen specified aircraft with certain clearly documented tail numbers each with 100 seats in them. The pilots' union stated if any of the 146s leaves the fleet they cannot be replaced – so Air Wisconsin can only operate a maximum of eighteen 146s and currently flies seventeen.

Shipping King Crabs with Air Pac

Alaskan regional Air Pac (or Air Pacific) ordered a 146-100 in May 1983 and N146AP was delivered on 3 March 1984. Prior to the 146's arrival, the fleet had only consisted of twin propeller airliners and smaller types.

Superlative airfield performance. Air Pac's N146AP (E1013) is taking off from the unsealed gravel strip at Dutch Harbour in the Aleutian Islands, off the coast of Alaska. (BAE Systems)

N146AP was operated on routes connecting the Aleutian Islands off the coast of North-West Canada and mainland Alaska, providing the first non-stop service between Dutch Harbour and Anchorage. At Dutch Harbour, the 146's performance was regularly tested on the unsealed 3,900ft crosswind strip, which runs like a channel through the side of a hill and ends in the sea.

Though not specially designed for mixed freight/passenger service like some of the later BAe 146s, in operations out of Dutch Harbour Air Pac's 146 often flew in a mixed freight/passenger combination with well-sealed containers of King Crabs in the forward section and the passengers in the aft section. Though this was a route network in which the 146 was in its element, the airline was not financially sound and ceased operations in 1986.

Aspen Airways

Aspen Airways of Denver, Colorado served the well-known mountain ski resorts of Aspen, Colorado, Colorado Springs and Durango, with other centres in New Mexico, Texas, Utah, Wyoming and Iowa with a fleet of Convair 580 turbo-props. It ordered the 146 in 1984 and received three between December 1984 and December 1986.

Dan Gurney, BAe Test Pilot recalled, 'I delivered N461AP (E1015) and supervised the entry into service The FAA could not believe that the 146 could follow the same flight procedures as the Convair turbo-prop and still climb out of the valley. Although the approach to Aspen is steep, it did not require short approach certification as a normal landing flare could be made. The only changes to the Aspen aircraft were the TACAN approach aid and clearance to lower flaps at 20,000ft.'

Aspen was taken over by United Express in 1988 and the 146s were disposed of in 1990. The first aircraft went on to serve with Paukn Air, Debonair and now flies for BA CitiExpress. The second flies with BAE Systems as a corporate transport registered G-BLRA and the final machine is with Air Wisconsin.

Presidential Airlines

Washington-based Presidential Airlines ordered five 146-200s with options on seven more in June 1986, which was especially welcome as it was the first substantial order that year. Since Hatfield had several unallotted aircraft, the first two aircraft were delivered in August and the others followed in quick succession.

In 1987, Presidential was contracted by Continental Airlines to act as a feeder from twelve East Coast cities. So in March 1987, to cater for this, Presidential decided to replace its 737s with 146s and ordered ten more 146-200s. However, events overtook this development and only the first three of the repeat order were delivered as Presidential was in trouble. In November 1989, British Aerospace grounded the carrier's aircraft after Presidential failed to make a $170,000 payment due and filed for protection under the US bankruptcy code.

Some of these 146s were refurbished for Discovery Airlines of Hawaii, which ordered five 200s and seven 300s, but only four were delivered before that airline also collapsed.

The 146 on the West Coast of the USA

The PSA Experience

After the success of the order from Air Wisconsin, Hatfield's sales team worked hard to procure another order from a large regional carrier in the United States. Byron Miller, Pacific Southwest Airlines (PSA) Vice-President, visited the 1983 Paris Air Show and was struck by the quietness of the 146. This immediately made it a strong candidate for the Californian operator, which had a noise-critical public at important points in its route network. PSA approached the 146 from a totally different perspective to Air Wisconsin by 'trading down' from larger, noisier Boeing 727s. The airline sought a competitive advantage by increasing frequencies on the lucrative internal Californian and Las Vegas routes at off-peak times when passengers are too few for McDonnell Douglas MD-80s, let alone 727s, to make a profit.

The critical factor in choosing the 146 was whether it could operate from John Wayne Orange County airport, which had eight noise-measuring stations plus an additional, and even more troublesome, form of noise monitor in the shape of a very sensitive retired colonel. Residents in the vicinity of the airport had placed a restriction on operations but would permit double the number of flights if the airline demonstrated that its aircraft did not produce more than 89.5dBA (decibels). And if the aircraft's noise did not exceed 86dBA then there were no operational restrictions. Even at maximum take off weight (MTOW) the BAe 146-200 was below 89.5dBA, which meant that the aircraft could fly to places as far as Seattle or Kansas City with a full load. On shorter sectors where less fuel was needed, the aircraft operated at a lower weight, could correspondingly use a lower power rating and so flights could be increased without any restriction. In contrast, the Boeing 737-300 could only stay below the noise limit with a severe weight penalty.

PSA was a tough customer, and needed a lot of convincing about the robustness of the powerplant before it eventually placed an order for twenty BAe 146-200s with twenty-five options on 12 November 1983. With first deliveries due in June 1984, only seven months after the order, this indicated the availability of plenty of production line slots before orders built up. The new customer insisted on new overhead lockers far larger than the existing 146 standard but similar to the Boeing 747 style. This was very different to the original concept and changed the whole feel of spaciousness in the cabin. It did however provide the customers with exactly what they wanted and these became the production standard until the late 1980s.

G-OPSA
Following on from G-SSHH's (E1002) activities during the test programme, it was refurbished at the BAe plant at Chester and re-registered as G-OPSA in honour of PSA. It left Hatfield on 17 March 1984 and toured the USA for sixteen days, visiting many places including the Avco Lycoming plant at Bridgeport where the engine was manufactured and Avco Aerostructures at Nashville where the wing was made. In Chicago it flew into Meigs Field, the waterfront airstrip adjacent to the city business area. With less than 4,000ft of runway, Meigs has no pretensions about jet airliner operations, normally only seeing propeller transports. On departing Meigs, Test Pilot Dan Gurney climbed the 146 to the height of the nearby 1,000ft Sears Tower some three seconds faster than the express lift takes to get to the top – and that was from brake release, not rotation! From Chicago it toured far and wide – Dan Gurney also demonstrated it in Florida to 'Air America' (a CIA-controlled 'airline' which carried out clandestine operations, especially in the Vietnam War) and there were many questions about infrared signatures and noise levels. Finally it was loaned to PSA for two months at their base at San Diego for crew training. During this period, it made 1,391 landings in 623 flying hours, converting sixty-four pilots.

First PSA Deliveries
The first two PSA aircraft were handed over together at Hatfield on 13 June 1984. As PSA was the official airline of Disneyland, local Hatfield school children were invited to the ceremony and Mickey Mouse and Donald Duck were on hand. Branded as 'Smiliners', each of the aircraft had a smile painted under the nose and the first two – N346PS and N347PS were named 'The Smile of PSA' and 'The Smile of Tri-Cities' respectively. After delivery across the Atlantic, the first services began on 27 June 1984.

From Six Abreast to Five
However, soon afterwards, problems began. The 146s configured in a six-abreast 100-seat layout with a single aisle. This did not suit the generally well-nourished Americans and passenger numbers dropped as passengers voted with their feet. PSA had no alternative other than to reduce the seating by fifteen seats to a more spacious five-abreast arrangement. This reduction might have adversely

Two of PSA's fleet, N349PS and N347PS (E2023, 25), at Los Angeles. Both were later stored in the Mojave Desert, and currently fly with the UK operator Flightline. (BAE Systems)

affected the finances of the whole operation, but the public were impressed, passenger loads recovered and its competitor Air Cal had to order six BAe 146s to compete.

The initial PSA order for twenty aircraft was delivered between mid-1984 and the end of the following year. Four options were taken up in 1986 and these were delivered in June/July 1987, by which time the aircraft were delivering a utilisation of ten and a half hours per day.

USAir Takes Over PSA

In 1987, PSA was bought by USAir, a large East Coast regional operator with a strongly unionised workforce. Pay rates for the 146 crews therefore increased, resulting in higher fares – substantially reducing the airline's competitive advantage.

As part of the rebranding of its diverse fleet, USAir repainted the 146s in the airline's livery which was polished bare metal with a red cheat line. When BAe was asked about this, the manufacturer warned them not to pursue this. However, USAir thought they knew better and removed the paint and primer to achieve a highly polished surface, which looked impressive. In so doing, seven

USAir's N188US (E2047) in the livery adopted when USAir took over PSA in 1988. USAir damaged the fuselages of seven 146s during the removal of paint and primer to achieve a polished surface, lowering their fatigue lives. (BAE Systems)

of the 146s had their fuselages damaged as the abrasion led to a thinning of the fuselage skin, which BAe asserted could become susceptible to cracking. As a result, these aircraft had a lower fatigue life so special checks had to be carried out on them at regular intervals.

USAir Ends 146 Operations

At the end of 1990, USAir announced a loss of $454 million and urgently set about cutting costs, routes and aircraft. So, in February 1991, the airline decided to sell their eighteen remaining 146s (one had crashed and five others had already been sold). In a press release accompanying this action, USAir claimed the reason was the aircraft's high operating costs. British Aerospace was stung to reply, citing that the 146 had been pivotal to the PSA route structure, the reduction in seating capacity was to improve comfort, the aircraft had proved themselves with despatch reliability of 99.1 per cent and that finally over-competition had resulted in a fares price war reducing revenues.

Parked in the Mojave Desert

The aircraft were withdrawn from service and parked in the dry atmosphere of the Mojave Desert, which is where many US airliners are parked, possibly awaiting resuscitation. The airline owned half of its fleet outright, the remainder belonging to different banks. Sadly, photos of serried ranks of cocooned 146s were no help to the aircraft or its reputation.

After four and a half years, the first of the 146s to escape the desert was E2047, formerly N364PS and later N188US, which routed through Calgary, Thunder Bay, Goose Bay and Keflavik to the UK, where it was refurbished at the Marshall factory in Cambridge. After 10,000 man hours, it was ready to enter service with UK airline Flightline as G-OZRH. Gradually, the other 146s were reactivated, though the final one remained in the desert for six years.

Air Cal

Air Cal was not slow to see the potential of the British jet and ordered six 200s at the beginning of 1986: all delivered between March and October that year. However the 146s were not long in use before American Airlines purchased Air Cal in July 1987. Like USAir, American Airlines has a livery based on polished metal upper and lower surfaces and wanted to treat the 146 in this manner. However, more wisely than USAir, they heeded BAe's advice and painted the aircraft an overall grey as the base colour.

With American Airlines large and diverse fleet, the six 146s were a very small item to integrate into the large technical and spares inventory. Serviceability began to suffer and so the aircraft were unable to even achieve 90 per cent despatch reliability. BAe had to lend another 146 to help – this briefly improved matters but in November 1990 the aircraft were returned to British Aerospace.

In 1991 the entire USAir BAe 146 fleet was withdrawn from use and stored in the Mojave Desert. N178US (E2040), formerly N357PS with PSA, with some of its compatriots. This aircraft remained in store for more than five years before re-entering service with the UK operator Debonair, and now flies in Australia with Quantaslink. (MAP)

Two former Air Cal aircraft with titling amended to American Airlines after the airline's takeover in July 1987. (BAE Systems)

WestAir

The third Californian airline to use the 146 was WestAir/United Express, which ordered six 200s with three options in November 1987. Hatfield was able to draw on three 'white tail' (unsold) aircraft, which had been in store at Woodford and quickly readied them for delivery. As a result the first three aircraft (N291UE - 3UE) were delivered only one month after the order in December, the balance following in mid-1988.

WestAir had a contract with United Airlines to run feeder services branded as United Express at the San Francisco and Los Angeles hubs. In December of 1989, United's East Coast partner (and 146 user) Presidential Airways went abruptly out of business. United called their WestAir partners, and asked if they would be willing to quickly assemble a new operation at Washington Dulles, which they did. Unfortunately, like Presidential before it, WestAir went out of business in 1993 and the 146s were returned to British Aerospace.

Canada

In 1988/89 two Canadian regionals, Air BC and Air Nova, purchased a total of ten 146-200s. Neither was able to buy more owing to scope clauses where the pilots' union restricted the number of aircraft bought with 80-100 seater capacity. These continued in service on domestic routes until they and other airlines were incorporated into Air Canada Jazz in 2002. Air Canada Jazz serves over seventy destinations in Canada and the United States, providing connections to the worldwide networks of Air Canada and the Star Alliance. The 146s, which are leased from BAE Systems Regional Aircraft, were all withdrawn from service by February 2005 and are being replaced by a mix of Bombardier and Embraer RJs.

A further Canadian operator was Halifax-based Air Atlantic, which served the north east of Canada and the USA. It received three former Air Wisconsin 200s between May and August 1990. These included the first 146-200 delivered (E2012), formerly Air Wisconsin's N601AW, which became C-FHAV. Air Atlantic ceased all flight operations on 24 October 1998 and its three BAe 146-200s were remarketed by BAe Aerospace Asset Management, which placed them with UK operator Debonair.

Air Atlantic of Canada leased E2066, formerly Presidential Airways N405XV, and seen here still carrying the American registration. It became C-FHNX. (BAE Systems)

Brazilian operator TABA's first 146-100 PT-LEP (E1010), delivered in December 1983, overflying Rio de Janeiro. (BAE Systems)

Brazil

Brazilian operator TABA (Transportes Aereas da Bacia), based at Belem, ordered two 146-100 on 17 March 1983 for delivery in December 1983. To clinch the deal, British Aerospace offered excellent terms of $40 million to be repaid over a ten -year period with no initial payment other than $2 million for spares. This customer was a perfect example of the BAe 146-100 customer just as the manufacturer had envisaged.

TABA's services reached into remote, inaccessible areas of the Amazon basin as well as flying feeder routes into and out of Belem and Manaus. The 146s would operate in an eighty-seat layout into short, unmade up airstrips replacing Fairchild FH227 turbo-props. The two 146s registered PT-LEP and PT-LEQ were delivered at the end of 1983 but proved an unfortunate experience for both the airline and the vendor. The airline complained about the after-sales service provided but the main problem was the high tariffs charged by Brazil on spares and TABA's refusal to pay them. As a result, through lack of spares one aircraft was grounded after two months and the other within a year. Both were eventually repossessed and returned to Hatfield.

Former BAe Test Pilot Dan Gurney recalled the colourful return flight of TABA's PT-LEP to Hatfield in November 1985. 'This was highlighted by the co-captain having little command of English and aircraft having only very restricted airframe systems. There was a thunderstorm in Curacao, a jail threat for the co-captain in Miami (no US visa), an engine failure in Bangor – so a three engine ferry to Stratford, Connecticut, for an engine change. The airframe anti-icing system failed in Greenland and that was followed by landing in a blizzard in Iceland. On arrival at Hatfield, the aircraft was fumigated with a "cyanide bomb", resulting in one dead snake and two large birds, one of which was in the air-conditioning.'

The two former TABA machines joined the second 146 (E1002) on lease to Royal West, which rather unsatisfactorily used them on the US West Coast, flying gamblers between Los Angeles and Las Vegas. But after just a year they were returned to BAe. The second 146 (E1002) then went to Birmingham, Alabama, for conversion to become the STA demonstrator.

Other South American Customers

The two TABA aircraft were not the only examples to grace the South American skies, as Lan Chile took over three of the former Presidential 200s in 1990. These flew on short-haul domestic routes to remote airfields high in the Andes and long intermediate flights to the USA. Lan Chile operated two of them for seven years after losing one through an accident in 1991. Aerosur of Bolivia also used two 100s and two 200s in the early 1990s.

Following its lease to Air Atlantic, E2066 was briefly used by AVIACSA as XA-RTI during 1991. It then served with various Canadian and German airlines before joining Club Air of Italy. (Author's collection)

Aviacsa of Mexico briefly operated three 146-200s in 1990-92 before replacing them with Fokker 100s, which were then replaced by Boeing 727s. Their operation was distinguished by an unusual situation in December 1991 recorded by Test Pilot Dan Gurney, 'The first real three-engine ferry was from Campeche (Mexico) to Little Rock via Brownsville with XA-RTI. A catastrophic engine failure had damaged the wing and adjacent pylon so much that an engine could not be carried. The flight was 'overland only', gear down, and unpressurised due to other damage, which had included a temporary wing main spar patch. When I first saw the aircraft, the engineers had mounted a cardboard missile on the unused engine pylon.'

Australia

Ansett Transport Industries announced a $40 million order for two BAe 146-200 aircraft plus spares with options on six more on 13 April 1984, for use by its subsidiary, Airlines of Western Australia. The first aircraft of the order, VH-JJP arrived in Perth on 26 April 1985 and the second 146-200, VH-JJQ, arrived on 26 June, just two days before services began between Kalgoorlie and Perth. VH-JJQ's delivery flight from Hatfield to Australia had taken only three days – Hatfield

VH-NJH 146-200 (E2178), delivered to Australian Air Express in October 1992. (BAE Systems)

to Larnaca, Cyprus (1,839 miles), in four hours forty-five minutes assisted by a tailwind, and then night stopped at Sharjah. From Sharjah it flew via Madras to Singapore for another night stop and then continued via Bali to Perth.

Though capable of carrying 100 passengers, these aircraft were configured more spaciously for seventy-five passengers at five abreast. Ansett continued purchasing BAe 146-200s until the early 1990s, augmenting Fokker F-28s on their Western Australian services. Later the same year they began servicing the Cocoas and Christmas Islands. Ansett Air Freight received its first BAe 146-200QT freighter, VH-JJY, on 8 May 1989. A second followed, and these replaced Fokker F27s.

East-West Airlines, also part of the Ansett Group, began operating BAe 146-300s in 1990, initially ordering eight, with options on four more. Their first two, VH-EWI and VH-EWJ, reached Tamworth, NSW on 29 August 1990, beginning services on 5 September. The combined Ansett/TNT group became the largest customer for the BAe 146 series. On 7 July 1991, the first BAe 146-100 for Australian Airlines subsidiary, Australian Airlink, arrived. Qantas bought Australian in 1992 and, from August 1993, Airlink BAe 146's began receiving Qantas colour schemes.

In 1996, Qantas subsidiary Southern Australia Airlines became a further BAe 146 operator, as did National Jet Systems when it took delivery of VH-JSF on 28 November 1995, for use on Indian Ocean routes from early 1996, as well as serving an Adelaide-Moomba contract with the SANTOS corporation. National Jet also took delivery of Australia's first Avro RJ, an RJ70 VH-NJT, on 8 December 1995.

Following the demise of Ansett in 2001, a large number of the fleet was put into store, some broken up for spares use and a few resold. Quantaslink continues to operate ten 146s; however, in 2005-6 all of them are due to be replaced by larger Boeing 717s.

New Zealand

At the Australian Bicentennial Air Show in October 1988 Ansett New Zealand announced an order for two 146-200s at $23 million each. They joined the airline in the following March, replacing Boeing 737s. They were soon joined by a further 200 and subsequently eight 300 series also featured on the New Zealand register while the 200s were sold. The British airliners served trunk routes linking the country's major cities and inaugurated jet services to several short airfields serving tourist areas. Ansett New Zealand later became Qantas New Zealand, which went bankrupt. Air New Zealand, which owned them, repossessed the aircraft and operated some for a time, but all are now stored in Australia.

China ordered ten 146-100s in mid-1985. This is B2709 (E1083) adjacent to the Hatfield Control Tower in July 1987. (Author's collection)

Far East Tour and Chinese Order

Following its American tour and PSA crew training, G-OPSA (E1002) returned to Chester on 17 June 1984 where it was readied for a five-week tour of the Far East. The main thrust of the 40,000nm tour was to demonstrate the aircraft in China, where the Hatfield team had strong connections owing to the thirty-three Hatfield-built Tridents supplied to CAAC, the state airline. Among the thirteen centres visited was Lhasa in Tibet (the highest airport in the world at 16,000ft) and the demonstrator flew fifteen sectors and made local demonstrations in China. The crew had to enjoy/endure seventeen banquets! Other countries included in the itinerary were Korea and Japan, where potential customers did three-engine handling trials into the sloping 3,600ft landing strip on Amami Oshima Island. G-OPSA then continued on to Brunei, Indonesia and Singapore before departing for the UK, visiting another high-altitude airport at Kathmandu, the capital of Nepal, on the homeward leg.

A demonstrable result of the tour was the visit of Chinese Premier Zhao Ziyang to Hatfield in June 1985 and an order for ten 146-100s as part of a massive re-equipment programme for CAAC. The first Chinese 146 flew from Hatfield on 7 April. Temporarily registered G-XIAN, it was handed over to CAAC on 2 September at the Farnborough Air Show and delivered on 10 September as B-2701. The remaining aircraft of the order were delivered over the following eleven months. The 146s operated into remote, sometimes unpaved airstrips often even lacking navigational aids.

CAAC was broken up into separate companies in 1989. Air China and China Eastern each received three 146-100s, while China North West had the other four and these were supplemented by eight stretched 146-300s delivered between 1992 and 1994. In 2002, China Eastern and China North West merged with a joint fleet of three 146-100s and seven 146-300s (one written off after a landing accident in 1993). The other 146-100s have been sold variously to British European, Albanian Airlines and to Moncrief Oil as a corporate jet. Some are in store and one has been undergoing trials as a fire-fighter for the US Forest Service.

In the Himalayas

Druk Air, the National Flag carrier of the Kingdom of Bhutan, commenced operations on 14 January 1983 with eighteen-seater Dornier 228s initially operating from Paro to Calcutta and Dhaka in Bangladesh.

Paro is the only airport in Bhutan and is located in a deep valley 7,300ft above sea level. The surrounding hills are as high as 16,000ft and approach into Paro airport is entirely by Visual Flight Rules. With the growth of traffic and the expansion of flights, the need for a larger aircraft became necessary. After conducting numerous tests, the BAe 146-100 was selected.

Before delivery to Druk, its first aircraft made the 146's third demonstration tour of Japan in March 1988. It was temporarily registered as G-BOEA and the twenty-two-day tour also included demonstrations in Hong Kong and Pakistan on the way home. On return to Hatfield, it was sent to Marshall of Cambridge for completion to Druk's specification and started operations as A5-RGD on 16 November 1988. A second series 100 followed in December 1992. With its introduction the network of Druk Air increased to link Paro with New Delhi, Bangkok and Kathmandu. In addition to its scheduled services, Druk Air's 146s also operate weekly sightseeing flights over the Himalayas from October to March.

During 2000-2001, corrosion was found in the wing tanks of A5-RGD, so its wings were replaced at Woodford and it was redelivered to Druk in August 2001. To cover for its absence, an Air Baltic RJ70 was wet-leased to enable the airline to maintain schedules.

Druk Air placed the first order for the Avro RJX development of the BAe 146/RJ in April 2000, with an order for two Avro RJX 85s. The RJX 85s would have allowed the airline to operate non-stop services to Bangkok (currently served via Calcutta or Dhaka) and Hong Kong. With the cancellation of the RJX in 2001, Druk Air was faced with a problem and ordered two Airbus A319s to replace its 146s.

One of the 146-100s of the Bhutan flag carrier Druk Airways in a typical Himalayan setting. (BAE Systems)

Other Asian Operators

The first sale of the 300 series to the Asian market was made when Thai Airways ordered two in January 1989 and later increased the order to five. British Aerospace obliged by leasing a 100 and five 300s to Thai, which were returned as five EFIS–equipped 300s were received. The 300s operated on domestic routes but were replaced with 737s in 1998.

Following the Taiwanese government's airline deregulation policy in 1987, new carriers were allowed to take part in both domestic and international operations. One of these was Makung International, which ordered five 146-300s, receiving the first aircraft B-1775 in August 1990. These flew in a high-density layout on domestic services, which had previously been the preserve of turbo-props. Their last aircraft, B-1781, first flew in July 1991 as G-BTUY and was used for trials with the new LF507 engine destined for the RJ. It was then stored at Filton and was the last 'new' 146 to be delivered in January 1995. Makung International was later acquired by Evergreen Group and changed its name to UNI AIR in March 1996. In April 1999, the 146s were returned to BAe after UNI AIR decided to replace them with additional McDonnell Douglas MD-90s. All now serve with other carriers except for the first machine, which was scrapped at Woodford owing to serious corrosion.

Initial Problems with the Avco Lycoming ALF502

The choice of the powerplant had been a major problem for Hawker Siddeley, but this had seemed solved when the firm alighted on the Avco Lycoming ALF502 with its phenomenal quietness – even though four engines were needed to power the aircraft.

With development occurring at reasonable pace, both BAe and Avco Lycoming believed it would prove reliable, but until then the engine had never seen intensive airline service and its initial reliability proved below expectations.

Difficulties with the engine were exacerbated by the poor performance of the starters, so an intensive programme was introduced to replace them. But the problems were dealt with, operating costs were driven down and reliability improved.

Early operators of the 146 tended to ignore the modularity of the Avco Lycoming ALF502 and take too much advantage of its ease of removal and replacement, which did not need to take more than an hour. Pacific Airmotive, PSA's maintenance company, generally employed this approach, but though quick and convenient it worked against the aircraft's economics, for the charge for removal was greater than the replacement of a part.

Demonstrating the ease of maintenance on the Avro Lycoming ALF502 engines. (BAE Systems)

But the engine was remarkably quiet and there was no obvious alternative to it. So its quietness was a selling point, just as the aircraft's four engine ability to climb to 20,000ft even after an engine failure, is far superior to a twin-engined aircraft after it has lost 50 per cent of its power.

Dan Gurney, 146/RJ Project Pilot, commenting on the engine's performance said, 'I had the first engine failure during performance trials. Whilst at FL300 (30,000ft) with one engine shut down, to measure drift down (we actually climbed on three), another engine suffered a rear bearing failure. This test confirmed the relight envelope at 20,000ft and the three engine handling landing which was a non-event; three-engine operations are essentially identical to normal operations. The engine performance deteriorated with high air bleed – airframe anti-icing, but again if the crew operated sensibly then there were few problems.'

8

The Royal and VIP Connections

Both heads of government and private operators have recognised the host of features the BAe 146/Avro RJ offers as a business jet to meet their specific VIP or corporate transport needs.

The 146/RJ combines outstanding airfield performance and reliability with an airliner-size cabin, allows access to the most difficult and remote airfields and provides the government or corporate operator with complete flexibility and independence. The large cabin – over three times the size of most large business jets – allows great flexibility for the user.

In Royal Service

Royal flying in Britain was formalised on 21 July 1936 with the creation of The King's Flight at Hendon. Initially using a twin-engine de Havilland Dragon Rapide, this was replaced in May 1937 by an Airspeed Envoy. After a break during the Second World War, in 1946, the King's Flight was reformed at RAF Benson with four Vickers Vikings. In 1964 Hawker Siddeley Andovers were introduced and saw more than twenty-five years of service before being superseded in 1986 by the 146.

Prince Philip made his first flight on the aircraft in Japan during the Far East Tour of G-ASCHH during November 1982. After observing from the jump seat, he took over for most of a flight between Tokyo Haneda and Sapporo Chitose Airport. John Nott, Secretary of State for Defence, announced in December 1982 that the RAF would purchase two 146s in the spring of 1983. These would be evaluated over a two-year period and, if this proved satisfactory, then the Queen's Flight would be re-equipped with the aircraft.

On return from route-proving, the fourth 146, G-OBAF, and the Far East Tour aircraft G-ASCHH were delivered to No.10 at Brize Norton. They were designated as BAe 146 CC Mk1 and laid out as an eighty-seater staff/VIP transport. Registered ZD695 and ZD696 respectively, the former served for

Handover of ZD696 (E1005) for twelve months' evaluation by the RAF on 17 June 1983 to examine the type's suitability for The Queen's Flight. E1005 had previously flown the Far East Tour as G-SCHH. (BAE Systems)

After the RAF evaluation, two 146-100s were ordered for The Queen's Flight. The second aircraft, ZE701 (E1029), was delivered in July 1986, and is now operated by No.32 (The Royal) Squadron. Just aft of the rear door is the Matador Electronic Counter Measures missile defence, now superseded by DIRCM. (BAE Systems)

eighteen months and the latter for approximately a year. They were comparatively under-utilised, flying only 800 hours during that period.

Following the evaluation, two new BAe 146-100 CC Mk2s were ordered with luxury cabins and wing-root fuel tanks to extend the range to 1,700 miles. The first aircraft ZE700 was flown from Hatfield in November 1984 straight to Chester for a lengthy fitting out.

ZE700 was only ceremonially handed over on 23 April 1986 at Hatfield when British Aerospace Chairman Sir Aubrey Pearce presented the logbook to the Captain of the Queen's Flight, Air Vice-Marshal John Severne. The second machine, ZE701, was also fitted out at Chester and followed in July 1986. Some may recall the use of ZE700 to take the Duke and Duchess of York on honeymoon – in fact the first royal use of the aircraft. As the aircraft taxied out to take off from Heathrow to the Azores, the rear airbrake petals opened to display the sign *Just Married*.

To add to the fleet, a third 146 joined the other two at RAF Benson on 21 December 1990. Ordered the year before, ZE702 was delivered to Marshall of Cambridge, which painted it and fitted it out to the rtoyal executive specification. This third aircraft caused a slight problem when it was being fitted out. The earlier duo had the initial smaller 146 overhead bins but by this time a much larger bin had superseded these. For sake of commonality the RAF insisted on the small bins so these were obtained from the Mali government 146, which was refitted with the newer, more capacious overhead lockers.

Royal Pilots

Prince Philip was a pilot for many years, often flying the BAe 146 and making his last flight from Carlisle to Islay in 1997. Until 1994 the 146 was regularly flown by Prince Charles who has a Private Pilot's Licence and was awarded his wings at the RAF College Cranwell in 1971. In November 1986, the prince undertook a two-day conversion course at RAF Benson, Oxfordshire, to enable him to take the controls of the BAe 146-100 four-engine jets, which replaced the propeller-driven Andovers in the Queen's Flight in July 1985.

Incident at Islay

Islay has had many visitors over the years and included many well-known and famous faces. Probably the most famous recent visitor is the Prince of Wales, who has made many visits to the island along with other members of the Royal Family.

On 29 June 1994 on a visit to one of his favourite Islay Distilleries, Prince Charles was piloting ZE700 when he encountered difficulties on landing. His personal pilot, the aircraft's captain, apparently failed to mention that the aircraft was landing with a tailwind and not into the wind. The 146 crossed the threshold

ZE700 (E102) at the end of the runway at Islay off the west coast of Scotland on 29 June 1994, after a landing incident when Prince Charles was at the controls, which damaged the aircraft quite seriously, but fortunately injured no one. (Callum Smith)

30 knots too fast and touched down on the nosewheel with approximately only half the 5,000ft runway remaining. Both pilots employed maximum wheel braking and deployed the lift dumpers but one of the mainwheels burst and another was partly deflated. Eventually ZE700 came to a stop more off the runway than on and damaging the 146 quite seriously. Luckily no one was hurt. The aircraft had, however, to stay on Islay until 18 August to allow it to be repaired. As a result, the Prince of Wales decided that he would no longer pilot royal aircraft.

No.32 (The Royal) Squadron

In 1995, The Queen's Flight was amalgamated with No.32 Squadron, which had previously operated Ministerial and Military VIP services, and was renamed No.32 (The Royal) Squadron. This rationalisation resulted in a better utilisation of the fleet and was centred on RAF Northolt just north west of London. Previously the Queen's Flight aircraft had positioned from Benson to either Heathrow or Northolt to collect their passengers.

The Squadron's current strength is two BAe 146s CC2, which carry nineteen to twenty-three passengers, five twin-engined BAe125s that carry seven passengers, and three Squirrel helicopters. In fact, royal flying accounts for only approximately 10 per cent of the combined tasking of both the BAe 146 and the 125s, which are more commonly used by senior military officers and Government ministers. About 60 per cent is ministerial and 30 per cent military and for foreign dignitaries.

Until recently there were three 146s, but ZE702 was withdrawn from use at the end of 2001 and stored at Northolt. It departed there for Southend as G-CBXY in November 2002. In August 2003, as PK-OSP it left on delivery for Indonesia for executive use by Surya Paloh, the owner of Indonesian Metro TV.

Princess Diana

After the tragic death of the Princess of Wales in Paris, ZE702 of the Royal Squadron transported her body to the United Kingdom on the evening of Sunday, 31 August 1997. The Prince of Wales and the princess's elder sisters, Lady Sarah McCorquodale and Lady Jane Fellowes, accompanied the princess's coffin on its return journey. Upon arrival at RAF Northolt, the coffin, draped with a Royal Standard, was removed from the aircraft's rear freight hold and transferred to a waiting hearse by a bearer party from The Queen's Colour Squadron of the RAF. The Prime Minister was among those in the reception party.

In order that the Squadron is well prepared for such an eventuality, the crew and ground staff regularly rehearse the transporting of a coffin of a member of the Royal Family.

Typical Sorties

The 146s of Number 32 Squadron travel widely. For instance, one of their pilots reported how he had visited sixty different countries and flown into 200 different airfields over a two-year period.

For a recent royal tour of the West Indies by the Duke of York in March 2004, the 146 had flown via Spain, Tenerife, Dakar, Conakry, Ascension, Recife and Belem to Barbados. The Duke of York flew out with British Airways, and on his arrival the 146 was there ready for him. Typically the aircraft would have two or three pilots, two stewards, four engineers and four Snowdrops (i.e. RAF policeman with their white-topped helmets) on board.

There are also spectacular sorties, such as flying into Kabul at night, using a steep approach with the aircraft's external lights out after a 3½ hours flight from Muscat. Then departing Kabul and arriving over Pakistan with no clearance, no radar service or TCAS at that time. This was flying VFR (Visual Flight Rules) at night and if other aircraft had their lights out there would be no way to avoid them.

Occasionally, the aircraft used to land on gravel strips, and the crew trained in Iceland and Canada for this eventuality. However, the 146's ability to land on short, critical runways is frequently put to use at airports such as London City or Plymouth. In the dry, the 146 can even land on Plymouth's 3,000ft runway at Maximum Landing Weight.

Stansted Emergency

On 6 November 1997, one of No.32 Squadron's 146s was forced to make an emergency landing at Stansted Airport with three of its four engines shut down. The incident occurred when two of the Squadron's captains, Squadron Leaders Jim Baker and Mickey James, were on a training flight from RAF Northolt to Norwich. Fortunately, no one was injured, and both the RAF and Serco, the private maintenance contractors, immediately launched internal inquiries.

Fifteen minutes into flight the oil gauges for engines numbers 2, 3 and 4 indicated empty and the gauge for the engine number 1 indicated less than one-quarter full. The crew began flying the aircraft back to Northolt. The low oil pressure warning light for the number 3 engine then illuminated. A little later the crew shut down the engine, declared an emergency and requested – and received – immediate clearance to land at Stansted. The low oil pressure lights for the number 2 engine and the number 4 engine then began to illuminate intermittently. The crew conducted an instrument landing system approach to Stansted with the thrust levers for the number 2 engine and the number 4 engine at flight idle, and the thrust lever for the number 1 engine at maximum thrust. When the crew was sure that the aircraft was in position for a safe landing, they shut down the number 2 engine. They shut down the number 4 engine during the landing roll and taxied clear of the runway using the number 1 engine.

On parking, the crew found the engine cowls were covered with oil, and oil spilled onto the ground when the cowls were opened. The engines hold 24.2 pints of oil; three engines needed twenty pints of oil to refill them, the fourth twelve pints. A leaked MoD report severely criticised Serco for poor maintenance standards.

The engine failures were determined to be the result of an engineer failing to fit oil seals when replacing the magnetic chip detectors fitted to each engine. As a result of their outstanding skill in landing the stricken aircraft, both pilots were, uniquely, awarded the Air Force Cross by the Queen.

Anti-missile Defence

Matador electronic counter measures (ECM) were originally fitted to the Flight's 146s in 1987-88 after identification of a potential IRA threat. In June 1999 an updated form of ECM, Directional infrared counter measures (DIRCM), as installed in ZE700, which became the RAF's prototype installation of this equipment. DIRCM will protect aircraft against modern infrared ground-launched missiles.

This system has four ultra-violet sensors to identify a threat, and to counter that threat it has one forward-facing dorsal nacelle well behind the nosewheel and two rearward-facing nacelles below and to the rear of the aft doors to

The Royal and VIP Connections

give complete cover. On the flightdeck there is a small black box loaded with software with all the decodes for the attack missile systems. If activated, the pilot hears a tone over his headset. Pilots do not manoeuvre the aircraft, as the system would then have to work harder. However, the pilot might reduce power to lessen the heat source that the missile is homing in on. Though the engines are the primary source of heat, other sources include pitot heaters and heated windscreens.

DIRCM operates completely autonomously using its own missile warning sensors, which identify the missile's rocket plume. It shines a gyro-stabilised infra-red beam onto the missile, confusing it so that it makes wild manoeuvres, it loses energy, slows and falls out of the sky. The system is readily upgradeable to meet the continually changing and emerging threats faced by modern aircrews and the infrared will soon be replaced with a laser beam system.

DIRCM confers a weight penalty on the aircraft. As the royal 146's carry a huge forward galley it was hoped that the DIRCAM would balance it, but there is a 3 per cent fuel-burn penalty at normal cruise speed.

With the increasing frequency of the Middle East as a destination, DIRCM is being enhanced and the aircraft had SATCOM (Satellite Communications) secure telephone, fax and web fitted in 2003, and SIFF/TCAS, a secure means for transponding needed for Iraq was fitted by Raytheon at Chester in 2004. As an added but rather more everyday form of security the aircraft's tails and wings are being painted white instead of red in 2004 to make them less conspicuous.

Other VIP Operators

UK

BAE Systems internal airline, Corporate Air Travel (CAT), operates two examples which both work hard carrying nearly 80,000 passengers in 2003. Based at Warton, CAT provides the company with point-to-point air travel, often directly into BAE Systems or partner company sites. The CAT fleet consists of a 146-100 G-BLRA, 146-200 G-TBAE and a BAe 125. To make the most of the investment, BAE sometimes charters the 146s out. Aspen Airways, Aerosur and Tristar previously operated G-BLRA, which flies with a sixty-six seater layout while G-TBAE is fitted with ninety-seven seats and previously flew with Air Wisconsin among other airlines.

The other UK executive operator is Formula One Administration, which uses two 146-100s in the corporate shuttle role to transport its television crews around the European Grand Prix circuits. They both carry branded registrations, G-OFOA (Formula One Administration), formerly with the RAF, Dan-Air and Australian carriers, and G-OFOM (Formula One Management) formerly in service in Indonesia and the United States.

In a rather less usual corporate use, BAe 146s played an important role in maintaining the smooth operation of the democratic process during the 1992 British General Election as both main parties operated examples. The British Conservative Party chartered former Loganair 146-200 G-OLCB and the Labour Party used British Air Ferries G-BTIA.

Mali Government

The Mali government was the earliest user of the 146-100 in this role, receiving TZ-ADT in 1983. Designed for use by the President, government officials and the national airline, it was the first 146 to be fitted with the optional wing root tanks to extend the range. A quick-change interior was fitted to alter the layout to accommodate dignitaries or airline passengers as needed. It was used on the President's visits to the people, calling at such evocatively named historic settlements as Timbuktu.

After just two years' service, it was repossessed by BAe and was stored at the company's Chester plant for five years. Following refurbishment at Fields Aircraft Services at East Midlands Airport, it now flies with National Jet Systems Qantas in Australia.

Both the Zimbabwe Government and Air Zimbabwe operated 146-200 Z-WPD (E2065). Owing to lack of spares, this aircraft was withdrawn from use in May 2000. (BAE Systems)

The Royal and VIP Connections

The Abu Dhabi government operated A6-RHK (E1091) from 1988 to 1999, but in the following year it was broken up at Basle. In this photograph, the fuselage is being loaded into Antonov An-124 for transport to Bournemouth Airport and scrapping. (D. Thomsen)

Demonstrating to Mugabe

Following G-OBAF's 1983 African Tour, there was another demonstration tour of Africa in late 1985 by G-SSHH. After demonstrations in Kenya, it flew on to Harare in an effort to win a Zimbabwean governmental order. As this was an executive sale, BAe wanted to show President Mugabe what the aircraft would look like in that arrangement and so eight rows of seats were removed and temporarily replaced with very plush hotel furniture from a Harare Hotel. Mugabe came on board and inspected the aircraft and agreed to wait five minutes to watch the 146 fly past. However, as the furniture was not fixed in place the pilots had to ask the two borrowed British Airways stewards to lash furniture down with tape – which proved successful. After the brief flight, the furniture was off-loaded and returned. BAe secured the order.

Following fitting out at Chester, Z-WPD, a BAe146-200, was delivered to the Zimbabwean Ministry of Defence as a VIP aircraft in January 1988. Owing to insufficient funds to order additional aircraft, the government regularly shared the 146 with Air Zimbabwe. This sole BAe 146 suffered from spares shortages and was withdrawn from use in 2000 because of the economic situation within Zimbabwe.

Abu Dhabi Government

Though first flown at Hatfield on 21 December 1987, as G-5-091, this 146-100 was stored at Cambridge until the middle of the following year when it was flown to Jet Aviation at Basle as G-BOMA for fitting out. It was finally delivered to the Abu Dhabi government almost a year after its first flight on 20 December 1988 as A6-SHK. It served with the private flight of the government for eleven years until flown back to Basle in 1999, where it languished until August 2000 when it was broken up because of wing tank corrosion, having only flown 3,251 hours. The fuselage was flown to Bournemouth in an Antonov An-124 and scrapped at Alton.

Pelita Air Services Indonesia

Pelita Air Services, a subsidiary of the Indonesian national oil company Pertamina, operates company support and VIP aircraft on the government's behalf. A BAe 146-200 'Statesman' PK-PJP fitted with pannier tanks was acquired in 1986 and fitted out at Chester with VIP layout in the front and passenger seats at the rear. Dan Gurney delivered it from Hatfield via Larnaca, Sharjah, Madras and Singapore, at which point BAe was paid for it and the aircraft was then flown on to Indonesia. Dan then made the maiden visit of a jet (the 146) into Solo with the President's wife, Madam Suharto, on board.

In late 1993, an Avro RJ85, PK-PJJ, joined it. As part of the government budget cuts forced by the East Asian financial crisis that began in 1997, the 146 was sold. As mentioned above, the third RAF No.32 Squadron machine also now serves in Indonesia registered as PK-OSP.

Other Operators

There is also one BAe 146-100, N861MC, which operates in a purely corporate role with Moncrief Oil of Texas. It is named 'Lucky Liz' and previously flew with CAAC, BAe and Carib Air. The Uzbek government has a VIP-configured Avro RJ85 UK 80001, which was delivered on 19 December 1997 but which flies in the Uzbekistan Airways livery.

Avro Business Jet

As part of a plan to further market the RJ for the corporate market, following its success as a 146 with the RAF and as an RJ with Pelita and Uzbekistan, BAE developed the Avro RJ Business Jet. To promote it, an Avro RJ100 (E3386) was furnished in an executive and corporate configuration. It first flew as G-NBAA in August 2001 and set off from Woodford on 11 September, heading for the National Business Aviation Association (NBAA) convention in New Orleans. With the closure of US airspace, it only got as far as Goose Bay, Canada and returned to Woodford on 13 September. The NBAA convention was cancelled. This frustrated demonstrator was later delivered to Blue1 as OH-SAM.

Future Developments

BAE Systems Regional Aircraft, mindful of earlier examples of the aircraft coming on the market, has identified a number of completion centres in North America and Europe that can convert aircraft to corporate configurations. It has also introduced a low utilisation maintenance programme geared to corporate aircraft operators where annual usage is typically 400-500 hours a year. It would seem likely that in the years ahead more of the quiet British jets will serve in the corporate/executive role, replacing noisier examples.

9

Quiet Trading

The British Aerospace 146 had many features that made and continue to make it a good choice as a freighter. It has a high wing, capacious fuselage floor, is low to the ground and above all is quiet – so important for much airfreighting which often takes place at night when airports are even more sensitive to promoting their 'good neighbour' policies. In earlier years, freight aircraft had often been noisy, gas-guzzling, elderly types, demoted to freight work from passenger flights. But such policies were becoming less sustainable and freight companies were having to look for more suitable machines to promote their services and brand.

Before finalising the design on a rear side-loading door, BAe investigated the installation of a cargo door in the front fuselage of both the 100 and 200 series. Owing to the restricted space available, the fuselage for the shorter 100 could only accommodate a 96 x 78in door to allow clearance from the engine nacelles. For the longer 200 series, a wider 140 x 78in door was planned. The cargo door for each would also incorporate the forward passenger door. In the event, BAe decided on a large cargo door fitted in the rear, aft of the wing, as this would allow for ease of access and could utilise a cargo door of common size for all the three 146 fuselage lengths. This version of the 146 was branded as the 146QT (Quiet Trader) with two versions – a 146-200QT and a 146-300QT on offer.

The Quiet Trader's Features:
- The 200QT carries up to nine LD3 cargo containers and the 300QT ten
- The 200QT takes six and a half standard 108 x 88in pallets, the 300QT takes seven and a half
- It can accommodate 125 x 96in pallets as well as long one-piece loads and can also transport bloodstock
- The 200QT cargo floor is 52ft 9in long, 10ft 7in wide and can carry a 26,000lb payload; the 146-300QT can take 27,500lb

- The hydraulic, infinitely variable, upward-opening freight door is 10ft 11in wide, 6ft 4in high with a sill height of 6ft 4in
- The flight deck occupants are protected from any dangers by a bulkhead; a smoke detection system and a facility to isolate themselves from the cargo deck air supply

TNT's £1 billion order

Having identified a market niche, British Aerospace converted an aircraft 'on spec' in order to promote the aircraft capabilities. The first aircraft for freighter conversion (E2056) left Hatfield as N146FT on 27 March 1986 en route to Hayes International Corporation at Dothan, Alabama, USA, where the conversion work was carried out. It returned to the UK five months later as N146QT in Quiet Trader livery.

Established in Australia in 1946, TNT provides global express, mail and logistics services and BAe made N146QT briefly available to them for two nights in November 1986 when it flew on several European routes. Following this trial, TNT recognised the QT's versatility for night operation out of noise-sensitive airfields, bought the prototype outright and placed an order for two more aircraft.

N146FT (E2056) at Hatfield prior to conversion to the prototype Quiet Trader (QT) at Hayes International, Dothan, Alabama. The position of the freight door is outlined in black on the rear fuselage. In the background is Air Wisconsin 146-200 N608AW (E2049), temporarily registered G-5-002. (BAE Systems)

1 On 3 September 1981, G-SSSH, the BAe 146 prototype (E1001), touches down at Hatfield after its first flight, while G-BFAN, its BAe 125 chase aircraft, overshoots. (BAE Systems)

2 The first 146-200 (E2008) on its maiden flight on 1 August 1982. It was registered G-WISC and painted in the livery of Air Wisconsin – the 146-200's first and loyal operator which continues to operate a large fleet. (BAE Systems)

3 Delivered as N461AP to Aspen Airways in December 1984, this 146-100 (E1015) now operates with British Airways CitiExpress as G-MABR. (BAE Systems)

4 Air Cal's N148AC (E2058), temporarily registered as G-ECAL for the 1986 Farnborough Show, together with Presidential N401XV (E2059) on the Hatfield apron in August 1986. (BAE Systems)

5 Air Wisconsin's BAe 146-200 N291UE (E2084) in United Express colours. Air Wisconsin is franchised by United Airlines to operate regional services. (BAE Systems)

6 ZE701, No.32 (The Royal) Squadron 146 CC2 (E1029) after being repainted in the Squadron's new toned-down livery in October 2004. Note the shrouded Electronic Counter Measure's nacelles under the front and rear fuselage. (Willem Honders)

7 Formula One operates two corporate 146-100s. This is G-OFOA (E1006) landing at Saanen in Switzerland, with airbrakes and lift dumpers deployed. This aircraft was formerly Dan-Air's G-BKMN, which flew the first 146 commercial service in 1983. (Markus Herzig)

8 The second of the five BAe 146-200QCs built, G-PRIN (E2148), at Hatfield. It was delivered to the short-lived UK operator Princess Air, and now flies with Titan Airways as G-ZAPK. (BAE Systems)

9 The Short Tactical Airlifter demonstrator, G-BSTA (E1002), in company with BAe Hawk 200 ZG200, at the Australian Bicentennial Air Show in 1988. (BAE Systems)

10 The 146-100 prototype G-SSSH was stretched in 1986-87 to become the 146-300 prototype G-LUXE (E3001). (BAE Systems)

11 The first UK operator of the 146-300 was Air UK. G-UKID (E3157) later flew with Buzz and now operates with CityJet. (BAE Systems)

12 Departing Innsbruck on 5 February 2005 is 146-200 OY-RCB (E2094) of the Faeroe Islands operator, Atlantic Airways. (Andreas Stoeckl)

13 East-West Airlines' 146-300 VH-EWI (E3171) and 146-200QT VH-JJZ (E2114), which, though in TNT livery, were operated by Ansett Air Freight. The former now flies with BA CitiExpress and the latter for Titan. (BAE Systems)

14 Ansett New Zealand's BAe 146-300 ZK-NZJ (E3147) 'City of Nelson', delivered to the airline on 19 February 1990. (BAE Systems)

15 One of Makung International's five 146-300s, which were operated by the Taiwanese airline between 1990 and 1999. (BAE Systems)

16 Thai Airways 146-200 HS-TBQ (E2074) and 146-300 HS-TBL (E3181) at Bangkok Airport in 1989. The 200 now flies with CityJet, and the 300 for Flybe. (BAE Systems)

17 In 1990 British Aerospace hired Aspen Airway's N463AP (E1063) for demonstrations as an RJ70 in the USA, and temporarily reregistered it as N70NA. The RJ development was not formally launched until June 1992. (BAE Systems)

18 A trio of development RJs in September 1992: G-BUFI RJ70, G-ISEE RJ85, and G-OIII RJ100 (E1229, E2208, and E3221 respectively). (BAE Systems)

19 In September 1993 RJ100 TC-THC 'Kayseri' (E3236), destined for THY together with RJ70 (E1223), was flying as G-6-223, and was later delivered to Business Express as N832BE. (BAE Systems)

20 Four of Sam Colombia's order for nine RJ100s in the Woodford flight shed in mid-1994. (Ian Lowe)

21 In celebration of Sabena's seventy-fifth anniversary in 1998, RJ100 OO-DWD (E3324) was painted in a special colour scheme – the only aircraft in Sabena's large fleet to be so distinguished. (Ian Lowe)

22 Woodford deliveries in November 1998: Crossair's RJ100 HB-IYZ (E3338) now operated by Swiss, Sabena RJ100 OO-DWG (E3336), now serving with SN Brussels and Northwest Mesaba RJ85 N518XJ (E2337). (Ian Lowe)

23 Park Express RJ100 taking off from Woodford as G-6-341 in January 1999, with part of its Turkish registration, TC-RJA, in place. This contract was not finalised and the aircraft was delivered to Aegean Airlines. (Ian Lowe)

24 BAe 146-100 (E1010) G-JEAO of Jersey European (now Flybe) began operating services for Air France in August 1999. (BAE Systems)

25 In late 2003, Aer Lingus withdrew its 146-300s from service. EI-CLI (E3159), EI-CLG (E3131) and EI-CLH (E3146) were stored in the Mojave in the USA in July 2004, with their Aer Lingus titling removed. (Derek J. Hellman)

26 One of Flybe's large fleet of BAe 146s climbing out of Manchester. (Derek Ferguson)

27 BA CitiExpress RJ100 (E3380) G-CFAD, making a sprightly take-off from Innsbruck in December 2004. (Andreas Stoeckl)

28 Air Botnia RJ85 OH-SAJ (E2388). Air Botnia has now been renamed Blue 1. (Ian Lowe)

29 The RJX development aircraft, nearing completion at Woodford. The first to fly was G-ORJX, an RJX85 (E2376) on 28 April 2001. G-IRJX (E3378), an RJX100, flew on 29 September 2001. (BAE Systems)

30 G-ORJX made its first public demonstration at the Imperial War Museum's airfield at Duxford, Cambridgeshire, on 21 May 2001. To promote the RJX's major customer, the aircraft wore British European titles. (Derek Ferguson)

31 On 10 January 2002, the three RJXs took to the air for a photograph session. The first and third aircraft were together – G-ORJX and G-6-391, an RJX85 and RJX100 respectively. G-6-391 (E3391) would have been delivered to British European, but it was dismantled in 2004. (Derek Ferguson)

32 Following many years of testing G-LUXE, the 146-100, and later 146-300, prototype was converted to become an Atmospheric Research Aircraft equipped with a multitude of sensors and the capability to carry a test-crew of scientists. (David McIntosh)

After installation of the freight door, N146FT was re-registered N146QT, and was used as a BAe demonstrator before delivery to TNT. Here it is shown in service with TNT as G-TNTA, showing off its sizeable freight door. (BAE Systems)

On 5 May 1987, the first aircraft entered service with TNT (though operated by Air Foyle) as G-TNTA and still serves with the company but is now on the Spanish register as EC-HDH. As TNT was not then a European-owned firm, it employed European operators such as Air Foyle to fly its aircraft.

On 23 June 1987 a further order was announced which was potentially one of BAe's largest ever. In their announcement, TNT expressed a firm commitment to buy every 146 freighter built over the following five years, which represented a massive boost for the programme and was a decisive 'stamp of approval' for the 146 from the world's largest transport group. The company calculated that with the lower operating costs and advantages of commonality, a single type would offset the higher capital costs of new aircraft against the usual elderly freighter aircraft conversions. For example, the 146s immediately started to earn their keep: with their quietness, they could make unlimited movements, saving TNT 10 per cent on noise-based landing fees at German airports.

TNT anticipated that a substantial number of the seventy-two QTs planned for production over the five-year period would be required to cover a worldwide expansion of its overnight freighting network. The remainder of the aircraft would be available to other freight operators through TNT's specialist sales and leasing subsidiary, Ansett Worldwide Aviation Services (AWAS), with Ansett Airlines – another TNT subsidiary – offering engineering and training support. The value of this order was deemed to be approximately £1 billion and covered both the 146-200QT and the 300 QT.

But the TNT contract was not a walkover for BAe, as its competitors, Boeing and McDonnell Douglas, were offering their 737s and MD-80s at lower prices than the 146. Eventually, British Aerospace clinched the deal by agreeing to sell the aircraft to TNT at cost.

TNT Express

In 1992, the Post Offices of Canada, France, Germany, Netherlands and Sweden bought 50 per cent of TNT, forming GD Net BV, though the company continued to trade as TNT. The new company revised its fleet strategy and so, for all the talk of seventy-two freighters, actual numbers delivered to TNT were only nineteen and total QT/QC production was only twenty-eight. This rethink was driven by effective hush-kitting of older types, enabling them to continue to operate, a substantial growth in traffic which led to a demand for larger types and higher than estimated operating costs of the 146.

A further ownership change took place in 1996 when the Netherlands Post Office bought TNT in its entirety after managing to divest the firm of Ansett and AWAS. TNT now carries over 187 million consignments in a typical year.

Ten TNT 146QTs at TNT's briefly used Cologne base. TNT's centre of operations is now Liège, from where it operates a fleet of twenty-one BAe 146 QT/QCs. (BAE Systems)

TNT Express employs more than 43,000 staff and operates forty-three aircraft, which offer daily connections between fifty-five airports in Europe, the USA and Africa.

Since entering service with TNT in 1987, the BAe 146-200QTs and 300QTs have provided quiet, reliable, nightly operations across Europe linking the company's hub, first at Nuremberg, then Cologne and, from 1998, at Liège – the most advanced air express parcels sorting centre in Europe. One of their points of call in the UK is Luton, and when the Luton runway was resurfaced between October 1988 and March 1989, TNT used nearby Hatfield for their services.

With its change in ownership, TNT Express became European-owned and established two wholly-owned airlines, the largest of which is TNT Airways based in Liège and PanAir, located in Spain. Between them, they operate seventeen BAe 146QTs (nine Series 200s and eight Series 300s). A further two 146QTs are operated by Mistral Air in Italy on domestic routes on behalf of TNT Express. In addition, TPG NV, the Dutch parent company of TNT Express, leases two 146QC freighters to Axis Airways in Marseilles, France. These aircraft provide French domestic freight services on behalf of TNT Express. The twenty-one-strong BAe 146QT/QC fleet represents almost half TNT Express's total fleet of forty-three aircraft.

Thoroughbred Airlifter

At Hatfield, on 10 March 1988, British Aerospace and TNT showed off G-TNTB to the aviation and horse racing press in a horse-carrying role. The 146-200 was fitted with four pairs of stalls for eight horses, which would be accompanied by an equivalent number of grooms. The larger 146-300 had the capacity for ten horses and grooms. Unlike some of the larger freighters, such as Douglas DC-8s which are used to transport bloodstock mares and foals, TNT targeted the market that flies a valuable racehorse from one event to another.

The features that sell the 146 to the passengers are also positive features for horses: its quietness, lack of reverse thrust on landing and good airfield performance, allowing it to operate into airfields nearer to race courses.

Australian Air Express

Two Series 300QTs and one Series 100QT are in service with Australian Air Express, which uses three aircraft operated by National Jet Systems on an overnight schedule linking Adelaide, Melbourne, Sydney and Brisbane, carrying 20 million kilograms of general freight, perishables and mail annually. The ultra-quiet operation of the BAe 146 is particularly important in this market and the aircraft is the only jet freighter allowed to operate during Sydney's stringent curfew hours.

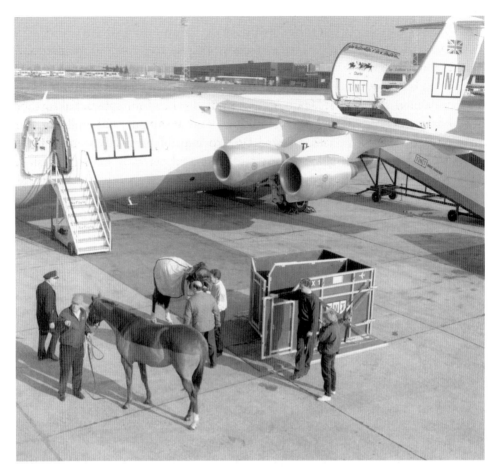

G-TNTE (E3150), a 146-300QT demonstrating its ability to transport horses which are loaded up the long gentle ramp into the spacious fuselage. (BAE Systems)

OO-TAK (E3150), a 146-300QT at Innsbruck in the current TNT livery. (Andreas Stoeckl)

The Opening of the Woodford Production Line

The TNT order underlined the announcement in March 1988 to open a second assembly line at Woodford to accelerate production to the forty aircraft per year that would be required to meet the expanding 146 airliner and freighter market. The availability of space at Woodford, without the need for substantial further capital investment (especially as £4 million had been put into building the new Hatfield Assembly Hall), allowed BAe to expand production above forty per annum should the market demand it. Sales had certainly been increasing, with nine aircraft delivered in June/July of 1987 in contrast to the mere twelve delivered throughout 1984.

There was another aspect to the opening of the Woodford production line. In his autobiography, Sir Raymond Lygo, BAe Chairman (and later TNT Chairman) wrote how the TNT order and the setting up of the Woodford production line was part of a strategy to close Hatfield, which finally happened in 1993.

Quiet Convertible

Building on the success of the Quiet Trader, a Quiet Convertible (QC) 146 was developed, employing the same freight door installation as the QT. However, the QC can be changed from a passenger to an all-freight or mixed layout in less than thirty minutes. The QC is easy to distinguish from the QT, as it has virtually the full complement of windows whereas the QT has none. As the QC's luggage lockers, toilets and galleys are not removed when the aircraft is utilised as a freighter the aircraft's capacity is slightly restricted in contrast to the QT and the 146-200QC can only carry six standard pallets in contrast to the 146-200QT's six and a half pallets.

The QC was announced at the 1989 Paris Air Show and BAe displayed a suitably modified G-BPBT (E2119) at the show. The go-ahead for the QC was based on fifty sales but unfortunately only five were ever built. The QC proved more expensive than the passenger 146 because of the installation of the freight door and reinforced floor.

QC Operators

Based at London Stansted, Titan Airways operates three 146-200 QCs and a single QT acquired in 2004 from Ansett Australia. The airline provides ad hoc passenger and freight charters for some forty European airlines, corporate charters and some work for the UK's Royal Mail. The three 146-200QCs can be chartered in a luxury seventy-seven leather-seater layout or can be reconfigured for freight in just thirty minutes. As mentioned above, the other two 146QCs are operated by the Axis on TNT freight services in France.

BAe's 146-200QC demonstrator G-BPBT (E2119), showing off its capabilities at Liverpool Airport. This QC currently operates with Titan as G-ZAPN. (BAE Systems)

QT/QC Production

Between 1986 and 1994, twenty-eight new-build BAe 146s were converted into the 146QT and 146QC variants. One further aircraft, a 146-100, was modified after several years' service. Production was as follows:

146-200 – thirteen QT freighters and five QC models
146-300 – ten QT freighters
146-100 – one aircraft, the second BAe 146 built (E1002), first flown in 1982, was converted in 1988 as the STA (Sideloading Tactical Airlifter) demonstrator and was civilianised in 1990.

All of the QTs and QCs are still in service today. Though the conversion proved successful, none of the later Avro RJs was ever converted into a freighter.

QT Future Prospects

With its noise compliance fully capable of Stage 4 noise limits, the existing 146QT/QCs are ideally placed to continue in service for years to come and, if demand grows, new conversions could be manufactured. BAE Systems Regional

Converting G-PRIN (E2148), a 146-200QC, to passenger configuration. Note the flooring and palletised seating. (BAE Systems)

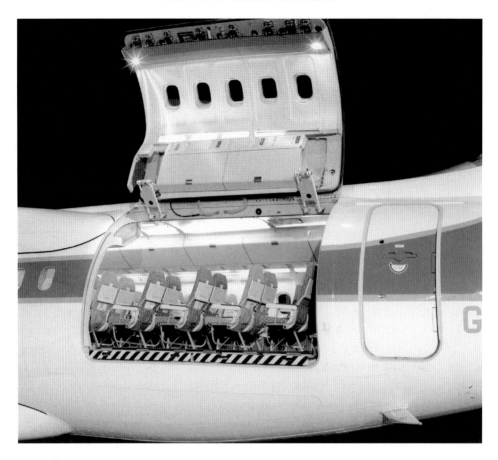

The seating fully installed in G-PRIN. Note the fully raised freight door containing the luggage lockers. This aircraft now flies with Titan as G-ZAPK. (BAE Systems)

Aircraft is actively assessing a new freighter conversion programme, as older 146s become more attractive to the freight market by virtue of their lower prices.

An initial ten-aircraft programme is being considered and, as the original QT contractor, Pemco, elected not to rejoin the project, a new conversion centre will be selected during 2005. If BAE decides to go ahead the first conversion should be fly during the following year.

10

Military Versions

Military developments of the HS 146 had been envisaged from the outset of the project and British Aerospace's relaunch of the 146 in 1978 was predicated on the sale of 100 military rear-loaders. In 1978, BAe stated that the military rear-loader was planned to fly in August 1982 but, owing to the many pressures on the company to get the airliner version into the air, the military versions were delayed and only first promoted in 1987.

Even if these military variants were only announced in 1987, a lot of work had been going on behind the scenes prior to this. BAe Woodford and the American firm Lockheed had been working on the design of a rear-loading Future International Military Airlifter (FIMA) and had identified the need for a small airlifter slotting in below Lockheed's own Hercules airlifter. A joint BAe/Lockheed market survey was carried out and, in May/June 1986, N802RW (E1010), flying in Royal West livery, was demonstrated to the United States Department of Defence prior to delivery to the airline. Lockheed and British Aerospace jointly organised the tour, which included the US Navy and Air Force test centres at Patuxent River and Edwards Air Force Base respectively. At Patuxent River, the 146 flew simulated carrier deck landings, which created a lot of enthusiasm amongst the Navy personnel.

Sadly, the success of these demonstrations caused an unfortunate rupture with Lockheed, which now viewed the 146 as a competitor rather than complementary to their Hercules, and they withdrew from the project.

Military Versions Announced

At the Paris Air Show in May 1987 four military versions were publicised, being the:
- BAe 146STA (Sideloading Tactical Airlifter)
- BAe 146MT (Military Tanker)

- BAe 146MSL (Military Side Loader)
- BAe 146MRL (Military Rear Loader)

Three of these military versions were based on the BAe 146QT Quiet Trader (QT) freighter and the similar Quiet Convertible (QC) which could carry freight or passengers or a combination of the two. Both the QT and QC were fitted with a large side-loading door. The odd one out was the Military Rear Loader, which required a major redesign and totally new rear fuselage.

STA Sideloading Tactical Airlifter

Only one of these types ever became hardware and this was the BAe 146STA. The STA version of the 146-200 series could carry a maximum payload of 30,000lb and featured the same size rear fuselage cargo door on the rear port side as the QT/QC. It had optional roller tracks to permit the movement of pallets and enhance the flexibility of the loading capability. The aircraft could land and unload standard military pallets via the cargo door, drop up to sixty fully equipped paratroops or carry up to twenty-four stretchers in the Casevac role.

MT & MRL Military Tanker and Military Side Loader

The Military Tanker was a 146-200 version of the STA with the large freight door, a flight-refuelling probe and two hose drum units mounted under the wingtips. The fuselage could carry four additional fuel tanks and to facilitate operations closed-circuit television would have been fitted. The aircraft would have had a range of 2,300 miles and been able to transfer 25,000lb of fuel. The Military Side Loader was a QT used in a military environment without any additional special equipment.

MRL Military Rear Loader

Producing a rear-loader version of the 146 seemed a sensible development of the type. Replacing the side-loading door with an air-openable ramped rear-loading door would enable small vehicles to be driven up the ramp into the cabin and enable dropping of them and supplies. A typical load could be two US Army 'Humvees' (HMMWV (High-Mobility Multipurpose Wheeled Vehicle)) or four pallets. In contrast, a C-130K could transport three such vehicles or five pallets.

The basic specification was as follows:
- A 100 series airframe with a new rear fuselage containing an air-openable, ramped, rear-loading door
- Lowered and strengthened floor
- Strengthened airframe and wings
- Tandem rather than dual-wheel main undercarriage

Military Versions

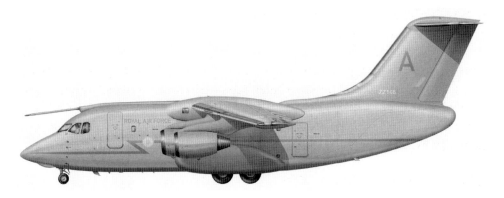

The 146 Military Rear Loader was based on the 100 series airframe with a new rear fuselage and air-openable rear-loading door to enable the transport and delivery of small vehicles and supplies. This is how it might have looked in RAF livery. (Rolando Ugolini)

- Larger nosewheel
- Double-hinged elevators
- Saddle tanks
- Flight-refuelling probe
- Navigator's station

If the MRL had gone into production, the aircraft would have been built and marketed at Woodford. It was logical that Woodford should take on the task of designing the rear loader, firstly, as Woodford was responsible for the manufacture of the rear fuselage and, secondly, because it was collaborating with European partners and Lockheed Georgia on the design of the rear-loading Future International Military Airlifter.

Naval Variants

Among many interesting proposed naval developments, which were never even publicly announced, was the 146N (Naval) Carrier on-board delivery (COD) project, essentially an STA designed for operation from large aircraft carriers. This project presumably grew from the success of the demonstrations at the USN's test centre at Patuxent River in 1986.

With its tremendous airfield performance, the 146 would have been a potent performer. The specification even examined below-deck hangarage – the extendable nosewheel would have been raised to allow the tail to clear the hangar roof and the folding wings would have also aided manoeuvrability and storage. Presumably, the carrier lifts were large enough and strong enough to cater for the aircraft.

The 146 (Naval) Carrier on-board delivery (COD) project was an STA designed for operation from United States Navy aircraft carriers. For manoeuvring on deck, the outer wings folded and the nosewheel could be extended to raise the nose for catapult take-offs. (Rolando Ugolini)

The extendable nosewheel would also have been used for catapult take-offs and the outer wings would have folded for deck operations. This sort of operation from an American aircraft carrier is not so implausible when reminded of the fact that Hercules have operated from such carriers.

The specification was as follows:
- 146-100 airframe with a freight door
- outer folding wings – outboard of the flaps
- strengthened airframe, wings and main undercarriage
- larger extendable nosewheel for catapult take offs
- arrester hook
- double-hinged powered elevators
- saddle tanks
- flight-refuelling probe
- navigator's station

Other Unpublicised Proposals

There were proposals for a mix and match of the various fuselage lengths and combinations of role including Naval Rear Loaders, AEW and Flight refuelling versions. These military versions could have taken advantage of the 146's various fuselage lengths depending upon payload/range/airfield performance criteria. Other high-tail freighter options included a version with twin Open Rotor Allison 578 contra-props and another with four General Electric TF34s.

The Military 146 Demonstrator – G-BSTA

To market the military versions the decision was made to produce a proof-of-concept demonstrator. So, the second airframe, a 146-100 which had initially flown on trials as G-SSHH, and in various guises with several airlines, was sent to

Military Versions

Hayes International in Birmingham, Alabama, for a freight-door conversion. It was fitted out by Hayes with a sixteen-seat forward cabin to demonstrate a military VIP role, a middle cabin demonstrating the Casevac and Paratroop role and a rear cabin fitted with a roller floor to facilitate pallet loading through the freight door. It first flew on 8 August 1988 and briefly returned to Hatfield on 11 August in glossy camouflage, appropriately registered G-BSTA. The same day it flew to Cranfield Aircraft Services where the aircraft was made ready for the Farnborough Airshow. An air-openable paratroop door and dummy, wooden, flight-refuelling probe were installed, while the camouflage was matted down from its inappropriate gloss.

STA Displays and Tours

G-BSTA was statically displayed at Farnborough, drawing a lot of attention from representatives of the World's Air Forces and Defence Organisations. Following Farnborough, the dummy probe was removed from G-BSTA, and the aircraft embarked on a thirty-six-day, 30,000-mile sales tour to Air Forces in Australia, New Zealand and South East Asia. Its first major stop was at the Australian Bicentennial Air Show at the RAAF base at Richmond, thirty miles north of Sydney, where it arrived on 9 October. During the following forty-eight hours, as the support team reinstalled the 'flight-refuelling probe', other BAe products arrived, namely a BAe 125-800, a two-seat Hawk 100, and a single seat Hawk 200. The Hawks were chasing a deal for the Australian Air Force trainer replacement (which BAe eventually won a decade or so later). An initial problem

The STA on show at Farnborough, 1988. (BAE Systems)

Paratroop dropping trials from G-BSTA in the winter of 1988/89. (BAE Systems)

for the BAe STA support team was that the side-loading door would not open, but after frantic telephone calls between Richmond, Hatfield and Dothan it was sorted out. After the show, the STA visited Canberra and Melbourne and flew into Puk-Puk, a 1,600ft unpaved airstrip. From Australia it flew over the Tasman Sea to Wellington for a day with the RNZAF and back via Australia to Indonesia, Brunei, Singapore and Thailand.

G-BSTA toured the Middle East in January/February 1989 and, on 6 May, left Hatfield on a tour of the USA with BAe Chief Test Pilot Peter Sedgwick at the controls. It was demonstrated to the Canadians first and then conducted some unusual and covert demonstrations for the US Army Special Airborne Forces, who did a parachute jump from it. It was a new experience for them to jump from a jet.

In May 1989, it made its first and only appearance at the Paris Air Show where the French national freefall team were very impressed by it and jumped onto the airfield at Le Bourget, only to be arrested for not having 'airside' passes.

The remainder of 1989 was taken up by a quick demonstration to the Norwegian Air Force and a number of appearances at UK airshows. G-BSTA was in Algeria in January 1990 but again no purchase was made. In the meantime, the Austrian Defence Ministry expressed an interest in replacing its Short Skyvans, and G-BSTA flew into Langenlebarn Air Base in April together with Hungarian-registered 146-200QT HA-TAB. The STA carried out its usual repertoire of short take-offs and landings, paratroop dropping and loading trials including the first landing on grass at Wiener Neustadt where the government and the UN flights were based. Though the 146 has often flown on gravel, it had never landed on grass. BAe had stipulated that the grass had to be dry – but it was wet and there were no problems.

Military Versions

Though BAe was disappointed with the lack of success, work was still being carried out to refine the product, so G-BSTA flew back to Dothan, Alabama, where the cabin was stripped and replaced with a complete freight-roller floor. Shortly afterwards it was displayed at Farnborough 1990, and in October a 146-200, G-BSRU, carried out tanker refuelling handling trials together with RAF Victor tanker XL190. These trials were carried out as part of a joint BAe/ETPS evaluation to assess if the 146 would be a stable flight-refuelling platform.

As BAe had achieved no sales during the two-year marketing programme the STA project was wound up. However, the Austrian government decided to purchase G-BSTA as a 100QC. It was positioned to East Midlands Airport where it was repainted in an attractive red and white livery and was provisionally allotted the registration OE-BRL and 'Republik Osterreich' titles. But, after these preparations, the sale was never finalised, and following service in South Africa this much-travelled and much-liveried aircraft is now a freighter with National Jet Systems in Australia.

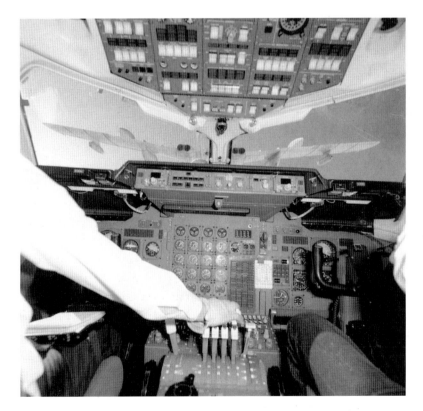

Spot the Handley Page Victor! For a joint BAe/ETPS assessment flight-refuelling proximity trials were carried out between 146-200 G-BSRU (E2018) and RAF Victor K2 XL190 in October 1990. (Note the pre-EFIS flightdeck. Compare with the flight deck photograph in Chapter 4.) (BAE Systems)

Interest in Military Versions

The STA inspired some interest, but, in the hostile theatre of war, customers wanted to be able to load or unload vehicles quickly without having to first erect a ramp and offload through a sideways door via a complicated series of manoeuvres. So, of the four military variants originally announced in 1987, the only real interest expressed was for the rear-loading version. However, without a definite order for twenty it was impossible for BAe to justify the £100 million investment.

The RAF were never interested in the project, so it remains something of a mystery as to how at the 146's relauch sales of 100 Military Rear Loaders were deemed a possibility. It is possible that if Lockheed had remained in partnership that it would have made some sales. But air forces were well served by the ubiquitous Lockheed Hercules and various other rear-loading types such as the Spanish-built CASA 212 and 235 military transports, which complement the Hercules. As a result, none of the military variants ever entered production.

In November 1990 the Austrian government indicated its intention to purchase G-BSTA. It was sent to Fields Aircraft Services at East Midlands Airport, repainted and registered as OE-BRL, but it was never delivered. It now flies with Australian Air Express. (Author's collection)

11

In and Out of the City

Richard Gooding, who has been Managing Director of London City Airport since 1996, said, 'The 146 would not have achieved the market it has without the Airport, and it is equally doubtful whether the Airport could have achieved its growth without the aircraft. We both needed each other.' The 146 was ideally suited for operations into a city-centre airport as it is quiet, has a high-drag landing configuration, powerful elevator control and good short-field performance.

The Development of London City Airport

The concept of a city-centre airport for London originated from a discussion, in September 1981, between Reg Ward, Chief Executive of the London Dockland Development Corporation (LDDC), and Sir Philip Beck, the Chairman of John Mowlem & Co., the large construction company with a long track record in the dockland. The LDDC was looking for an innovative project to shift perceptions of the Royal Docks and to improve accessibility to the area. An airport seemed to fit the bill and Sir Philip, a pilot himself, took the idea to Bill Bryce of Brymon Airways, the Plymouth-based airline which specialised in the use of Short-Take-Off and Landing (STOL) aircraft including the de Havilland Canada Dash-7. Within weeks, Mowlem and Brymon had developed an outline plan for an STOL port for the Royal Docks, and with LDDC support the project was launched. However, it proved to be a very long haul and it was nearly six years before the first passenger-carrying flights took place. Though services started in 1987, they proved very limited, but the following year the BAe 146 entered the equation and the prospects for London City Airport began to improve radically. The BAe 146-200 would be able to quickly and easily reach such major European centres as Copenhagen, Berlin, Stockholm and Zurich, an attractive idea for travellers, but it was one which would undoubtedly arouse a good deal of comment from anti-noise protesters.

The 146 Demonstrations

As a prelude to possible developments, John Mowlem & Co., as owner of the airport, and supported by the two airlines that operated from there – Brymon and London City Airways – arranged an open day on Sunday, 24 July 1988 to permit the public to see the 146 at close quarters. Under the original agreement, jet-powered airliners were not cleared to use the facility, so it was anticipated that there would be forceful opposition to any suggestion of a change. Considerable effort was therefore made to allay the fears held by the local inhabitants to assure them that noise levels would be low. In fact, there were very few banners or vocal chants as visitors arrived to view the various routine movements and await the arrival of a pair of 146s.

Whereas the de Havilland Canada Dash-7 operating into London City (LCY) typically flew a 7.5° approach, for the obstacle clearance, 5.5° was adequate. So, the BAe pilots concentrated on proving that the 146 could fly that approach very successfully on ILS, in all weathers. In preparation BAe set up a visual guidance system at Hatfield to show that the aircraft could fly the approach and a special steep ILS at BAe's airfield at Dunsfold, Surrey (more usually accustomed to Harrier test flying), to show that the autopilot could follow it. Commencing in January 1988 the first steep approach tests were flown with G-BMYE (E2008). The 146's versatility meant that little had to be changed for a steep approach. A steep approach requires less thrust, and to increase drag, airbrakes and full flap are deployed which allows the aircraft to come down much steeper, while control response remains the same and the excellent cockpit view enhances the situation.

BAe had rightly identified the significance of the event and how it might improve sales, so demonstrated their product in style. First Loganair's G-OLCB, a 146-200 captained by BAe Chief Test Pilot Peter Sedgwick, made the shallow 5.5° degree approach into London City, landed and parked in front of the crowd to give them a close-up look. The weather was ideal, because it was a very windy, gusty day and the film taken showed how well the aircraft handled in challenging conditions and how accurately the British Aerospace team had forecast the noise levels would be.

The second 146, company demonstrator G-BMYE with BAe's Dan Gurney at the controls, followed, landing effortlessly on the short strip. G-BMYE then made an equally impressive take-off, the length of the run being little more than that of the turbo-props. In order to judge the effects of noise from a fully laden 146, the aircraft was refuelled until the take-off weight was roughly equivalent to a seventy-passenger load. It did a series of take-offs and landings and flew around the local area to demonstrate to those at the airport and in the area what it would really be like. In these conditions it was to be expected that the noise level would be greater, but most of the locals likely to be affected seemed pleasantly surprised

The first landing of a BAe 146 at London City Airport on 24 July 1988 as part of an Open Day to allow the public to see and hear the aircraft at close quarters. G-OLCB (E2103) was in Loganair trim, but was only delivered to them in May 1989. (Derek Ferguson)

at the impressive display. A London City Airways Dash-7 then replicated the exact flight pattern, so onlookers could make a real comparison of the noise made by the two different types.

Protestor... 'could not hear the 146 land'

The objectors lost half their number after they heard the first aircraft land. The *Daily Mail* reported that, 'Resistance Leader Mrs Leisha Fullick... admitted she could not hear the 146 land'. A MORI opinion poll of 1,250 people after the event found that only one in six opposed the introduction of the 146.

Steep Approach Certification

The initial 146 Steep Approach Certification by the Civil Aviation Authority (CAA) was based on Canadian STOL rules: these became the basis for the Joint Airworthiness Authority (JAA) rule. The certification for 100 and 200 series allowed the aircraft to fly a steep approach even with one engine out.

Any 146 flying into LCY has to have the approved steep approach modification, which is essentially a switch that inhibits some of the Ground Proximity Warning System. The 300 series has more severe payload limitations under some circumstances, although these are not a factor on most of the short legs that the aircraft tend to operate. There is an increased possibility of a tail strike under certain conditions in the 300, as the fuselage geometry is different, but that is really a training issue rather than an aircraft limitation, and 300 series/ RJ100 operate regularly out of LCY.

G-OLCB, seen under the wing of a de Havilland Canada Dash 7. (BAE Systems)

Extending LCY

There was work to be done before the 146 could become a regular sight at LCY, much of it involving the extension of the runway to 3,934ft. Owing to various Governmental and bureaucratic hurdles, it was only on 15 December 1991 that the airport began operating under a new licence allowing aircraft to land on a 5.5° glidepath instead of the previous 7.5°.

In February 1992, a 146-200 operated by the Swiss regional carrier Crossair became the first foreign registered jetliner to visit LCY. The airline subsequently launched a twice-daily schedule between the docklands airport and Zurich on 30 March, pioneering regular jet operations at LCY. Response to this service was better than expected, although it took two years before the route reached the break-even point with just over 43,000 passengers carried in 1993. With traffic figures steadily improving, Crossair announced that it intended to offer another link with Switzerland from October 1994, this time to Geneva with through connections to Lugano.

More 146 Operators

During this growth period by the Swiss regional, LCY was chosen by another two 146 operators, the Swedish Malmö Aviation and the German airline, Conti-Flug. The former inaugurated a link with Stockholm-Bromma, another truly city-centre airport. Most of Bromma's commercial traffic had moved to the more distant Stockholm Arlanda some years before, but special dispensation was accorded the 146.

London City handled one million passengers in the first seven months of 1997 compared with 727,600 during the whole of 1996. This impressive landmark was reached by the combined efforts of the airlines then operating from LCY, many of which were flying 146/RJ to eighteen European destinations including Brussels (Sabena), Dublin (CityJet), Frankfurt (Lufthansa/Business Air), Geneva, Lugano and Zurich (Crossair), Malmö and Stockholm (Malmö Aviation), Milan, Rome and Turin (Azzurra Air), and Paris (Air France/CityJet).

Nevertheless, without doubt much of the growth was due to the ever-improving access to the docklands facility. When LCY opened in 1987, it was certainly not the easiest of destinations to reach, so the convenience of the uncongested terminal was cancelled out. This is no longer the case following the constant development of the area with journey times to the centre of London and elsewhere drastically reduced. In addition to the 146/RJ, the only other jets certificated for operation at LCY are the thirty-seven-seat Embraer 135 and the eighty-seat Fokker 70 – a shortened version of the Fokker 100.

In 2004 carriers serving London City with the 146/RJs were Air France to Dublin, Paris Orly and Charles De Gaulle, British Airways CitiExpress to Edinburgh, Frankfurt and Geneva, British European to Belfast City and the Isle of Man, CityJet to Dublin, Lufthansa to Frankfurt and Swissair to Basel, Geneva and Zurich. Notably, many of these were 'blue chip' carriers taking advantage of the British jet's outstanding performance.

Flying into London City – A Pilot's View

Chris Grainger, a CityJet pilot who regularly flies into London City, describes what it is like to operate the 146 into the airport.

The success of 146/RJ operations at London City Airport is indicated by these five airlines where 146/RJs are represented: Malmö, Lufthansa, Alitalia, Crossair and KLM UK. And there are many others. (BAE Systems)

'London City airport is the benchmark for most airliners. They can either manage the 5½ degree glide path, a mandatory touchdown zone and 3,600ft of concrete to stop in – or they cannot. The other criterion for anything that operates out of City is the climb gradient after take off assuming the loss of one engine.

The first few times into City are quite harrowing if you have become used to long wide runways, because it looks very short, and very narrow, and you are after all looking at it from twice the normal approach angle. After a few visits you realise that the aircraft will stop at maximum landing weight in less than 2,000ft, in other words about half the runway length. The take off at London City Airport also demands a similar distance.

This performance is due to the very large double slotted fowler flaps which cover almost two-thirds of the total span and deploy to 33 degrees for landing (30 degrees are available for a short take off). Additionally the lift spoiler system dumps the full weight of the aircraft onto the main wheels immediately after touchdown, supported by very effective anti-skid brakes, which might also be very helpful in the event of a rejected takeoff.

Controllability is also a major issue at City airport: as previously stated, the runway is narrow so whatever the crosswind (up to 20 knots) touchdown must take place on the centreline, and before the touchdown zone lights set 300m in from the touchdown end of the runway.

As the planning people have seen fit to allow the building of windshear generators in all directions nice and close to the runway confidence in the fact that the aircraft will point precisely where you want it to point at any given time is a great comfort. The combination of ailerons/roll spoilers and a very powerful rudder achieve this very nicely.

In normal flight the aircraft has a solid feel, gentle inputs resulting in gentle manoeuvres, more sporting inputs (with an empty aircraft as the passengers do not appreciate it) result in a startlingly quick rate of roll, and new converts to type often find themselves overcontrolling in roll, especially on approach.

A pilot's-eye view of London City Airport. It is actually taken from BAe Jetstream 41 G-JMAC during steep approach certification in June 1999. (Derek Ferguson)

Malmö Aviation operates a fleet of 146/RJs out of Stockholm Bromma, the capital's city-centre airport. (BAE Systems)

Stockholm-Bromma Airport

Stockholm-Bromma was built in 1936, only five miles from Stockholm city centre but it was outside the city limits. Over the intervening years the Swedish capital has spread to encompass Bromma and a new airport was opened at Arlanda, thirty miles from the centre, while Stockholm-Bromma was only used for corporate and general aviation. Then, in 1992, deregulated airlines began to use Bromma again and one million passengers now use it annually. Björn Rotsmann, General Secretary of the City-centre airports Association, said there were many reasons to use Bromma, 'Aircraft have become more environmentally friendly: BAe 146/RJs often go unnoticed by the local populace generally hostile to airport noise.' Malmö Aviation first introduced 146-200s in 1990, later replacing these with nine former SAM 112-seater RJ100s which are the largest aircraft permitted to operate from Bromma Airport, which has been its hub for many years.

Even though some other regional jets can now operate into these city-centre airports it is the British jet that made operations at London City and Bromma viable.

12

From Hatfield to Woodford

The Largest Write-off in British Commercial History

In 1992 BAe came close to ruin when it was hit by a financial crisis brought on by the potential financial exposure to its Regional Aircraft lease book. The number of aircraft on lease had expanded rapidly in the 1980s as the sales force had sought to place the 146s, ATPs and Jetstreams with customers.

Throughout the late 1980s, airlines found themselves in an increasingly precarious situation, with weakening capital bases. This was particularly acute in the regional sector. The deregulation of the US airline sector had led to a large number of start-up airlines, all with weak balance sheets but a considerable appetite for small aircraft. BAe, with its large production operation, needed to generate cash from sales. So in order to maintain sales the manufacturer was forced to offer leasing arrangements whereby financing houses provided the capital, while BAe guaranteed the lease arrangement. The manufacturer essentially sold aircraft to finance companies, which then leased them back to BAe for it to remarket them to airlines on short, flexible subleases. (For example, aircraft were placed on leases of three years or less while the manufacturer arranged a head lease for fifteen to eighteen years with the banks.)

This strategy seemed to work, but then in 1992 the civil market took a dramatic turn for the worse. Airlines began returning significant numbers of their leased aircraft at appreciably depleted values with little prospect of finding new customers at sustainable rates, leaving BAe to pay the lease rentals to the financiers. The impact on profit and cash was disastrous. Faced with this situation, BAe decided on a major restructuring of the whole Regional Aircraft business and made a special provision in its accounts of £1 billion. At the time, this was the largest corporate write-off in the history of the UK.

Financially the company was on the brink of disaster. Reserves were virtually wiped out and net worth was reduced to £1.7 billion. This was only £100 million above the lower limit set in the firm's banking covenants and if it had dipped beneath this threshold BAe would have been in the hands of the banks.

The BAe Regional Aircraft business had been driven by the need to fill the factories, rather than delivering valuable business. The need to keep moving aircraft off the production line and generate cash had meant that aircraft were over-financed in order to maximise the sales price. Customers' plans were not assessed or monitored and so, when aircraft were returned, BAe was not ready and the aircraft became a cost to the company.

So, with any continuing production, there had to be real sales not leases. The objective was to do business with blue chip operators, such as Lufthansa, Sabena, Swissair and BA. The intention was to go for quality rather than yield – better to take a lower price but get paid in cash.

British Aerospace decided on a dramatically reduced production rate, down from forty-two aircraft a year to eighteen to twenty per year. (Similar reductions were made in the ATP and Jetstream production.) Production would be limited and market driven not product driven. If BAe could not supply an aircraft quickly that was not to be a problem.

Asset Management Organisation

To deal with this large fleet of leased aircraft (which included ATPs and Jetstreams), in January 1993 BAe established the Asset Management Organisation at Bishop Square, Hatfield, to manage these aircraft, quite separately from the sales organisation. With better planning and more rigorous management they were quickly able to recover and stabilise the situation within a short period. This division between asset management and sales resolved any possible conflict for British Aerospace in marketing new RJs quite separately from used or new 'white tail' (undelivered) 146s.

In January 1993, British Aerospace had 118 BAe 146 aircraft on its books, twenty-one idle and forty due to be returned that year. At the peak of the recession there were forty-four BAe 146s parked. Within three years, asset management had reduced the idle fleet to zero and raised the market value of the lease rates to a level that supported the sales effort of new-build aircraft.

Actions taken included agreeing residual values for all the aircraft and then measuring and pricing the risk according to the strength of the airline, any other relevant factors and instigating ongoing credit. All contracts were reviewed and informed decisions made to balance and manage the risk for the whole portfolio.

Insuring the Risk

Later, BAe was able to break new ground by inventing a new form of insurance, and insured the future income receipts on virtually all the remaining leases the firm had guaranteed. During 1998, an external review was commissioned of the likely income to be generated from the portfolio of aircraft to which the company had such financing exposure. This review identified a probable income of £2.4 billion. Following this analysis, BAe entered negotiations with a syndicate of leading insurance companies and was successful in concluding underwriting arrangements covering £2.2 billion.

The Closure of Hatfield

Throughout the 1980s Hatfield's productivity continued to improve. Whereas in 1988 it took thirty-eight weeks to fully complete a 146 nose section, by January 1992 this was taking only fifteen weeks. In December 1990, a 146-300 (E3186) was assembled in just nine weeks. But the firm's clear strategy was to divest itself of its sites in the south – Hurn (1983), Weybridge (1987) and Kingston (1992) – and sell them where possible for redevelopment and concentrate in the north. It was announced in March 1991 that henceforth 146 final assembly would be concentrated at Woodford. A year later, on 23 March 1992, the first RJ85 (E2208) registered G-ISEE, and the last aircraft to be completed at Hatfield, made its first flight piloted by Dan Gurney. This was to be the last of 8,468 aircraft built at Hatfield since 1934.

In July 1992, the staff newsletter *Focus on Hatfield* was still speaking about the future of the plant, but on 23 September management announced the phased closure of Hatfield to the workforce. BAe 146/RJ nose and 300 series fuselage extension production continued at Hatfield until the beginning of October 1993, but after that work gradually wound down and the last BAe staff finished at the end of the year. On 4 April 1994 the famous airfield finally closed. The last aircraft to fly out of Hatfield was very appropriately a Tiger Moth, G-APLU. It was piloted by Dick Bishop, 146 Training Captain and son of the famous de Havilland designer R.E. Bishop together with Anne Essex, granddaughter of Sir Geoffrey de Havilland, in the passenger's seat.

BAe 146-NRA (New Regional Airliner)

When Charles Masefield became Hatfield Divisional Director and General Manager in April 1986 he asked Future Projects to investigate the possibility of replacing the four ALF502 fan engines on the 146 with two fuel-efficient fan engines. The 146-NRA was proposed, powered by two of the well-established CFM56 engines produced by CFM International, a firm jointly owned by the American General Electric and the French engine maker SNECMA. This engine had become the launch powerplant for the Airbus A320 family, the Boeing 737-300, and is fitted to all the later marks of the 737.

The BAe 146-NRA project. A lengthened, re-winged, 110-136-seater development, powered by twin CFM56 engines. (BAE Systems)

BAe had intensive discussions with CFM and agreed on the CFM56-F5 version for the 146-NRA which would be derated to increase times between overhauls and would not need a thrust reverser. Charles Masefield was then approached by Northwest Airlines, which wanted to replace their noisy Boeing 727s and Douglas DC-9s. The requirement was for a new airliner capable of carrying 120 mixed-class passengers over a range of 1,800 miles operating up to 35,000ft. This would put the airliner into a new league and bring it head-to-head with the Boeing 737-500.

A new large wing with greater sweep was required to meet increased weights and speed. Northwest required an EFIS (Electronic Flight Instrument System), which Boeing was not offering on the 737-500 while BAe was in the process of updating the 146 flightdeck to EFIS and so could offer it on the 146-NRA.

So, in 1991, before the crisis hit British Aerospace, the design of a radical new development of the 146, the twin-engined 146-NRA, was frozen and its maiden flight planned for early 1995 and first delivery in mid-1996.

BAe 146-NRA Specification in April 1991

- Able to carry 125 passengers 1,750 miles in a five abreast layout
- Powered by two 21,000lb thrust CFM56-F5 engines derated to 19,000lb
- 146 fuselage cross-section
- Fuselage length 119ft 3in
- Trimming tailplane (in contrast to the fixed tailplane on the existing 146 family)
- New 97ft 1in, 25° swept wing with winglets, leading-edge slats, kruger flaps and doubled-slotted, trailing-edge flaps
- Flat panel display flight deck and optional sidestick control
- Fy-by-wire controls

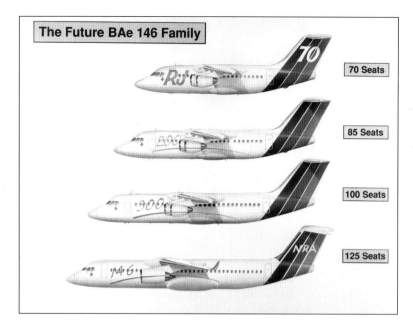

The proposed 146 family with the 146-NRA as foreseen in 1991. The 146-100 has been rebranded as the RJ70, while the 200 and 300 series remain unchanged. (BAE Systems)

BAe 146–NRA and Taiwan

However, before the project could proceed, BAe's financial crisis caused a major upheaval and a radical restructuring. BAe then entered into a long series of negotiations with Taiwan Aerospace (TAC) and the two sides signed a Memorandum of Understanding in September 1992, ratified in January 1993. Under the agreement, TAC was to buy a 50 per cent stake in BAe's Regional Aircraft operation, which was expected to cost TAC about £150 million, including its contribution to development and certification of the new 146/RJ Series.

The agreement called for the setting up of a 146/RJ production line in Taiwan and the transfer of some part production. However, these talks never came to fruition as the Taiwanese sought a definite go-ahead on the 146-NRA (now renamed the RJ-X) while it appears that BAe only wanted to establish shared production of the RJ family. Meanwhile, BAe's financial position improved after the sale of various assets and the sale for $372 million of Corporate Jets Division (the Hatfield-designed BAe 125 executive jet programme) to Raytheon in the USA. With these disposals Regional Aircraft was able to survive without a partner.

If the NRA had gone ahead it would have set a real challenge for the BAe marketers who would have been extolling the virtues of the four-engined RJs over the twin-engined competition and would then have had to praise the twin-engined layout of the NRA.

Rebranding – 146 to Regional Jet

Bombardier first coined the term 'Regional Jet' in 1987 when it announced the development of a 'regional jetliner' version of its Canadair executive jet, the fifty-seater Bombardier CRJ100. The first series of this jet, the Canadair Challenger 600, had been powered by the Avco Lycoming ALF502 turbofan, but in 1982 these were replaced by General Electric CF-34 powerplants.

The definition of a Regional Jet in its simplest form is a jet-powered airliner with no more than 100 seats. The term 'regional jet' conjures up a short-range aircraft, but many of them have a substantial range in the order of 2,000 miles.

Bombardier has continued to stretch its basic design from the CRJ100 up to CRJ900, which has capacity for eighty to ninety-nine passengers. Likewise, in 1999, Embraer of Brazil launched the Embraer 170, 175, 190 and 195 to build on its smaller jets as members of a new seventy-110 seat family of aircraft. These two programmes, heavily supported by each company's respective government, have continued to attract orders.

Sales Success

By 1990, with some 202 aircraft ordered and 157 delivered, British Aerospace could feel satisfied that the 146 had proved a success in competition with the Fokker 100. However, BAe was aware that there was a need to address misgivings amongst some in the airliner market. Engineers had never had any doubt about the sturdiness of the airframe but had reservations about the engine configuration. In producing a jet that could operate into marginal airports, BAe's engineers' parameters were limited. By choosing a four-engined layout, they secured an excellent engine out performance, but when all other types were twin-engined this was swimming against the tide of perception if not actuality. Where such performance was critical, the 146 sold well, for example to operators using London City Airport where the 146 was the only jet airliner certified to operate from it. Switzerland's regional carrier Crossair became a confirmed user because of the 146's ability to operate into confined alpine airports.

Major Improvements

Just prior to the 1990 Farnborough Air Display BAe announced a number of major improvements to the 146 family. The aircraft would continue to be available in the three existing fuselage sizes but would have:

- A new version of the Textron Lycoming ALF502 powerplant, the LF507 with Full Authority Digital Engine Control (FADEC)
- New flightdeck avionics with Category 3A autoland
- Reduced empty weights and higher operating weights to boost payload
- Increases in capacity, payload and range – payload up by 28 per cent (100 series), 13 per cent (200 series) and 12 per cent (300 series)
- A new 'Spaceliner' interior with better seating, lighting and luggage bins

The most important change to this new version of the 146 was the introduction of the LF507 engine. This change was intended to lower maintenance costs, which had proved an impediment to some sales.

Textron Lycoming carried out modifications to the ALF502, which improved performance, and later produced the LF507 with a higher thrust rating, an extra low compressor stage and FADEC which provided smoother running. The LF507 was a great improvement, with a removal rate that was half that of its predecessor.

Launch of the RJ Family

BAe toyed with the Regional Jet branding in 1990 and projected versions of the 146-100 called the RJ70 and RJ80 with de-rated engines and lower weights than the 146-100. An Aspen 146-100 was hired and demonstrated in the USA in 1990 as N70NA emblazoned with RJ70 branding, but it was only in June 1992 that the three new versions of the 146 family were formally launched. To emphasise the improvements these were rebranded as Regional Jets, so each model would be identified by their passenger capacity with five-abreast seating:

- RJ70 (former 146-100)
- RJ85 (former 146-200)
- RJ100 (former 146-300)

In the initial promotion of the RJ family there was also a version of the RJ100, which, with six-abreast seating and an extra emergency exit, was dubbed the RJ115, as it would have capacity for approximately 115 passengers. This photograph gives the impression that this version actually flew. (Compare this with the similar photograph in the colour section.) (BAE Systems)

For a time, there was also an RJ80, a six-abreast version of the RJ70, and an RJ115, a six-abreast version of the RJ100, which needed an additional mid-cabin emergency exit, was never certified. Later, to simplify the branding, the RJ80 and RJ115 designations were dropped, leaving just three main types.

As an additional marketing stimulus, the RJs were offered with customer support packages: for example, an Engine Maintenance Cost Protection Program (EMCPP), which guaranteed reliability, and maintenance costs would be no higher than a comparable twin-jet. This initiative took the unpredictability out of the maintenance costs. Most operators took advantage of this and other extensive warranties for their leasing and sales contracts with BAe to guarantee reliability and economy.

In November 1992, in order to broaden their product range, Fokker began development of a derivative of the Fokker 100, the shorter Fokker 70. The Fokker 70 had commonality with the Fokker 100 in its systems, Rolls-Royce Tay engines and EFIS flight deck. Like BAe, Fokker was after the forecast 2,000 plus aircraft in the seventy-125-seat market estimated as being required over the next twenty years.

RJ Test Programme

Prior to the go-ahead of the RJ and its new LF507 engines, three aircraft were employed to trial the replacement powerplant. The 146-300 prototype G-LUXE, 146-300 G-BTUY (E3203) and 146-200 G-BTVT (E2200) all spent various periods on engine trials in the USA during 1991 and 1992.

The development aircraft for the three RJ versions made their maiden flights in quick succession and embarked on a test programme at Woodford. The first development aircraft (and Hatfield's last aircraft), G-ISEE, an RJ85, made its maiden flight from Hatfield on 23 March 1992 with 146/RJ Project Pilot, Dan Gurney, at the controls. Dan said, 'On its maiden flight, we carried out two faultless Category 3 Autolands, one of them in a 14 knot-crosswind. In fact, the whole four and a half hour flight was fault free – remarkable for a new development aircraft.'

Dan Gurney also made maiden flights of the first RJ100 (G-OIII) on 13 May and the first RJ70 (G-BUFI) on 23 July 1992, both from Woodford. All three RJs made their maiden flights unbranded, but this was soon rectified and the RJs were painted in an attractive RJ livery. The trio demonstrated at Farnborough 1992 and a subsequent in-flight photo sortie over the Isle of Wight produced some stunning images of them.

The last Hatfield-assembled aircraft, G-ISEE (E2208), then unbranded as the RJ85 development aircraft waiting to taxi out for its first flight. It was the last of 8,468 aircraft assembled at the Hertfordshire factory between 1934 and 1992. (BAE Systems)

Rebranding – 146 to Regional Jet

The most important change between the 146 and the RJ was the replacement of the ALF502 powerplant with the digitally controlled LF507. This photograph shows the installation on G-OIII, the RJ100 trials aircraft. (BAE Systems)

The main emphasis of the RJ test programme was on proving the new avionics and improved engines, though all the performance trials were re-flown and altitude limits increased. Though the avionics were developed on the simulator and accepted by the CAA test pilot on the simulator, it was still felt that a simple flight test was necessary for certification.

The first production RJ, an RJ85 for Crossair temporarily registered G-CROS (E2226), made its first flight at Woodford on 27 November 1992 and immediately joined the test programme. Joint Airworthiness Authorities certification was granted to the RJ85 on 23 March, to the RJ100 on 22 July and finally to the smallest of the RJs, the RJ70, on 24 August 1993. Federal Aviation Administration certification was received on 3 September 1993. There was the same type rating for all of the 146 and RJ variants – so pilots could fly any of them.

Painted in Air Malta livery, G-OLXX (E1228) was used for a forty-four-day tour of the Middle and Far East in 1994. (Ian Lowe)

RJ Demonstrations

During February and March 1994, RJ70 (E1228), painted in Air Malta livery and appropriately registered G-OLXX (L + X + X = seventy in Roman numerals) made a forty-four-day sales tour of the Middle and Far East. During the 39,000-mile tour, it was demonstrated to twenty-six airlines. On return it continued in this role with BAe for over a year and was demonstrated to other operators before delivery to National Jet Systems in Australia as VH-NJT.

Avro International Aerospace

In 1993, British Aerospace's RJ division at Woodford was rebranded as Avro International Aerospace with a new logo and the RJs were now referred to as Avro RJs and sometimes as Avroliners.

Woodford Flight Test became Avro Test and led the testing for all the current BAe civil aircraft – the RJ, Jetstream 41, BAe ATP (later Jetstream 61) and the BAe 125-800 and 125-1000 executive jets.

14

The RJ in Service

With the granting of certification, the first deliveries of the RJs took place. Crossair of Switzerland received its first aircraft, now registered HB-IXF, on 23 April 1993. Likewise, THY, the Turkish national airline, received the first of five RJ100s, TC-THA, on 22 July and Business Express's first RJ70, N832BE, left on its delivery flight across the Atlantic on 9 September. The RJ soon acquired a reputable order book and British Aerospace was obviously proud of the fact that the RJs were ordered in the colours of 'blue chip' operators such as Lufthansa Cityline, Crossair, THY, Sabena and British Airways. (With the Turkish Airline, the RJ100 actually replaced a fleet of nine DC-9-30s.) These airlines were much more secure than many of the earlier regional operators that had ordered the 146 and whose weaker finances had contributed to BAe's financial problems in 1992.

RJ Performance Improvements
From 1996, weight savings, together with drag reduction improvements, were introduced as standard, which reduced fuel consumption by 5 to 10 per cent depending on variant and sector length. The cabin pressure differential was increased, allowing cruising altitude to increase by 4,000ft to 35,000ft.

Competition with Fokker and Boeing
British Aerospace was justifiably pleased that between 1993 and 1995 it had won orders from three of the five major European airlines in the market for regional airliners: Crossair/Swissair, Lufthansa and Sabena. Alitalia chose the Fokker 70 but only briefly operated it. (Alitalia contracted Azzurra Air to operate services with RJ70s and RJ85s on its behalf under the Alitalia Express brand, from April 1998 to March 2002.) Though BAe tried hard to win an order from SAS, demonstrating 146s to the airline on numerous occasions, SAS ordered Boeing

737-500s. The 737-500, which entered service in 1990, was the smallest member of the 'Second Generation' 737 family, but with a 130-seat capacity could not be regarded as a real competitor to the British airliner.

By 1995, the RJ was outselling its main rival – the Fokker 70 and 100 – by two-to-one. However, the Fokker 100 was particularly successful in its early years, obtaining large fleet orders from the USA, where American Airlines had a fleet of seventy-five and USAir a forty-strong fleet. As a result, the average fleet size of the Fokker was eight against four for the 146. However, the RJs redressed the balance and their average fleet size equalled the Fokker's at eight.

But, after its earlier successes, Fokker got itself into trouble by having a surfeit of 'white tails'; i.e. complete but unsold aircraft on their hands, and the Dutch aircraft manufacturer went into liquidation in 1996. When production ceased in early 1997, a total of 283 Fokker 100s and forty-eight Fokker 70s had been built.

In Service in Europe

Crossair's 'Jumbolino'

Between 1990 and 1993 the Swiss carrier Crossair operated three 146-200s that had previously been with assorted American carriers. Crossair was impressed with the 146's airfield performance – it was able to fly into London City and Swiss airports such as Berne and Lugano, and quietly, which was all very important in noise-conscious Switzerland. In 1992 Swissair acquired a majority stake in Crossair and in the same year it became the first airline offering jet flights to London City Airport. In 1993 the fleet of British airliners was expanded when

Swiss RJ100 HB-IXX (E3262) at Manchester Airport in May 2003. (Derek Ferguson)

The RJ in Service

D-AVRF (E2269), one of Lufthansa's eighteen-strong fleet of Avro RJ85s, touching down at Manchester in March 2003. (Derek Ferguson)

four RJ85s were delivered to replace the BAe 146-200s. The RJs like the BAe 146-200s before were branded as 'Jumbolino'. Two years later twelve RJ100s were ordered, to operate flights on behalf of Swissair, which sold its Fokker 100s to BAe Asset Management in part-exchange.

Swiss Air Lines, operating as Swiss, is the new Swiss national airline formed from Crossair and the remnants of Swissair. It formally began operations on 1 April 2002 and its fleet includes Crossair's four Avro RJ85s and fifteen RJ100s. In 2005 Lufthansa, a committed 146/RJ user, agreed to buy Swiss.

Air Malta

Air Malta chose to order four RJ70s after a tough battle with Fokker, who had promoted their Fokker 70 to the Mediterranean carrier. Air Malta's RJ70s were a higher weight, longer-range version of the RJ70. They had a higher MTOW and were fitted with optional wing fillet tanks and fully rated LF507s to enable them to fly Luqa-London with a full load. Air Malta later added three RJ85s to their fleet, leasing them all to the Italian operator Azzura Air, who in turn leased them to Alitalia, who operated them as Alitalia Express for five years.

Lufthansa CityLine

During mid-1994 Lufthansa was seeking a type between the fifty-seat Canadair RJ and 108-seat Boeing 737-500. They evaluated the RJ and visited Crossair to garner information first hand from a user rather than the manufacturer. It was a hard-fought battle between BAe and Fokker for the Lufthansa order, but BAe won and the German flag carrier ordered three RJ85s for delivery later that year – BAe taking Lufthansa's Fokker 50s in part-payment.

The first aircraft, appropriately registered D-AVRO, was delivered on 17 October. Because of a dispute with the pilots' union, it was painted in CityLine Europe colours. During introduction of the RJ85, there were 'Scope clause problems' as in the USA with the permitted upper limit of seats per aircraft with Lufthansa CityLine. Lufthansa pilots feared that CityLine would take over more Lufthansa services eroding their pay scales, so an agreement was reached that Lufthansa CityLine could only operate a maximum of eighteen aircraft with eighty seats. The subsequent three aircraft were also delivered in CityLine Europe colours until an agreement with the pilots' union allowed Lufthansa titles to be carried. At the end of the year Lufthansa CityLine gave the RJ a vote of confidence by ordering seven more RJ85s, and eventually they purchased a total of eighteen.

Stuttgart's Saviour

Stuttgart is Germany's sixth busiest airport with over five million passengers carried annually. During closure of Stuttgart's main runway for resurfacing in 1995, there was only a 5,000ft secondary runway available. BAe 146/RJs operating with Lufthansa and seven other airlines provided the vast majority of both scheduled and charter flights during this period, Lufthansa's Avro RJ85s were even used on the three-times-daily flights to London Heathrow.

The British jets could easily carry a full payload from Stuttgart to places as far as Heraklion on Crete. Normally only Sabena and Hamburg Airlines operate the type from Stuttgart but other airlines, LTU, Deutsche BA, British Airways and THY, all used the British jet while the main runway was out of action.

Lufthansa Regional – the World's Largest 146/RJ Operator

In October 2003 Lufthansa set up Lufthansa Regional, a branded airline grouping of five airlines – including Lufthansa Cityline, Eurowings, and Air Dolomiti of Italy. In early 2005, Eurowings leased six 146-300s to add to its existing fleet of eight and Air Dolomiti leased five BAe 146-300s. This 30 per cent increase in capacity makes the Lufthansa Regional grouping responsible for the world's largest BAe 146/Avro RJ fleet with thirty-seven aircraft. This was a powerful endorsement for the long-term market acceptability of the 146 with a blue-chip operator, underlining its competitiveness compared to newer regional jet equipment.

Best-selling British Jetliner

With the delivery to Lufthansa of D-AVRD, BAe's 250th 146/RJ, on 23 March 1995, the type had outsold the BAC One-Eleven of which 235 were built in the UK (and nine in Romania). (This figure does not include G-LUXE, the 100/300 series prototype, or G-BMYE, the 200 prototype.)

In the previous year, BAe Woodford had delivered seventeen RJs, including eight to SAM of Colombia and three each to THY, Lufthansa and Air Malta. This was very different to 1991 when the joint production lines at Woodford and Hatfield had produced thirty-nine. Even though production was at a lower rate than in the late 1980s and early 1990s, on 1 December 1996 the 300th aircraft flew. This was EI-CNJ delivered to Azzurra Air, which was delivered later the same month.

Sabena

The Belgian national airline, Sabena, placed the largest ever order for the 146/RJ with a contract for twenty-three RJ85s on 1 September 1995, narrowly beating the order for twenty from Pacific Southwest in 1983. These aircraft were to be used by its regional subsidiary Delta Air Transport (DAT), which already operated six 146-200s in Sabena livery. Initially, it was planned the BAe would take back all of the six 146-200s delivered to DAT, but in the event most of them still operate for DAT's successor, SN Brussels.

Importantly, this order was a cash deal rather than a lease agreement – very much one of the post-1992 objectives of BAe Regional Aircraft. The fact that Sabena's subsidiary already operated the 146 (it was the first national carrier to operate a 146 fleet) helped BAe, as did the 49.5 per cent shareholding Swissair had in the airline. Sabena chose to emulate Crossair by fitting its machines with eighty-two leather seats in a five-abreast layout.

In September 1996, Sabena exercised an option to convert part of its order for twenty-three RJ85s into RJ100s. The Belgian airline took delivery of the first of nine Avro RJ100s, OO-DWA on 27 June 1997. Like the RJ85s, the RJ100s were operated by Sabena's affiliate, Delta Air Transport.

After the terrorist attacks on America on 11 September 2001, many airlines suffered badly. Sabena was owed substantial sums by Swissair. After Swissair stopped operations on 2 October 2001 and refused to repay the money, Sabena was also forced to stop flying. Sabena failed to gain financial backing and went into liquidation on 6 November 2001. DAT took over its own operations and became SN Brussels. It continues to operate a fleet of six BAe 146-200s and twenty-six Avro RJ85/RJ100s, making it the third largest operator of the aircraft in the world and the second largest in Europe. In 2005 SN Brussels renegotiated its contracts with BAE on the RJ/146 fleet and the British jet will remain in operation with the Belgian airline well into the next decade.

British Airways CitiExpress

British Airways CitiExpress, a wholly owned subsidiary of British Airways, operates services from seventeen airports in the UK and Ireland on sixty-three routes to major or central regional airports in the UK and Europe. It was formed on 31 March 2002 through the merger of British Regional Airlines and Brymon. Manx Airlines, which had taken over Loganair and its two 146-200s, G-OLCA and G-OLCB, in 1994, joined the grouping later that year, giving the fleet a total of five BAe 146s – three 200s and one each of the 100 and 300 series. As part of a general restructuring of British Airways, by the end of March 2003 it had also received the sixteen former BA CityFlyer Express Avro RJ100s. Five of the RJ100s have received the modifications necessary to operate steep approaches for the routes from London City, which began on 30 March 2003 from London City to Frankfurt, Paris and Glasgow. The CitiExpress aircraft vary widely in age – from 146-100 G-MABR, which first flew in 1984, to RJ100 G-CFAH, which had its maiden flight in 2001.

Aegean Airlines

Established in 1999, Aegean Airlines introduced flights from Athens to Heraklion and Thessalonika in June 1999 using its two newly delivered RJ100s. Only four months later a further RJ100 joined the fleet as new destinations were inaugurated to Alexandroupolis, Kavala and Corfu. In December 1999 Aegean Airlines bought Air Greece and purchased another brand new RJ. Aegean's RJ100s are configured in a two-class, 100-seat layout and achieved a despatch reliability of 99.98 per cent and average load factors exceeded 75 per cent. So, in July 2000, two more RJ100s were added to the fleet. Aegean now flies to eleven cities in Greece with more than eighty daily flights.

Air Baltic and Air Botnia/Blue 1

Air Baltic, founded in 1995, operates to many European destinations from Riga, the capital of Latvia, and Vilnius, the capital of Lithuania. In January 1996, it leased three former Business Express Avro RJ70s, which were delivered between January and May. Air Baltic's RJ70s typically fly up to six daily round trips between Riga and Copenhagen. One of the RJ70s has spent much time on lease to other operators such as Portugalia, National Jet Systems of Australia, Azzura Air and Druk Air.

Whereas Air Baltic's main shareholders are the Latvian State and SAS, Air Botnia was founded in 1988 as a domestic Finnish airline, which was soon taken over by SAS. In March 1999, the airline's network strategy was changed. The new aim was to develop the airline into a regional carrier, feeding traffic between points in southern and western Finland and SAS hubs in Stockholm,

Aegean Airlines' first RJ100 SX-DVA (E3341) taking off on a production test flight from Woodford, temporarily registered as G-6-341. This aircraft originally flew in the colours of Turkey's Park Express but, when this order was not confirmed, the RJ was allotted to Aegean. (Ian Lowe)

Copenhagen and Oslo, as well as to operate frequencies and routes where the size of SAS-operated aircraft was not optimal. The first of the initial lease order for five British Aerospace RJ85s, OH-SAH, was delivered on 10 May 2001.

Air Botnia acquired the last four unsold RJs – two RJ100s and two RJ85s – on a long-term lease in 2003. These machines flew in 2001/2 and were stored at Woodford, then Filton, before completion by Flybe at Exeter and delivery. The leases on the earlier aircraft were extended at the time of this deal. Air Botnia changed its name to Blue1 on 1 January 2004 and the four RJs were delivered in the new livery during October and November 2003. As mentioned at the beginning of this book, their last aircraft, OH-SAP, had the distinction of being the final British airliner delivery.

In Service in the Americas

SAM Colombia
SAM Colombia was the first airline in the Americas to buy the RJ100, placing an order valued at $230 million for eight in June 1994. SAM was the regional subsidiary of Avianca and the RJs were used to replace the Boeing 727-100s

The 100th RJ taking off from Woodford's runway 07 on its maiden flight on 20 January 1988, carrying the pre-delivery registration G-6-321 (E3321). It was delivered to Northwest Mesaba on 30 January 1998 as N509XJ. (Ian Lowe)

operating on domestic and international routes. Between December 1994 and the following September, SAM received a total of nine aircraft, the first being originally the RJ100 development aircraft, G-OIII. Owing to the negative growth of the Colombian economy, all of the fleet were returned to BAe in 1999/2000. After overhaul they were leased to Malmö Aviation as part of a rollover scheme where most of its older 146-200s were returned to BAe for lease or sale. Malmö's hub is Stockholm's city-centre airport Bromma (see above). Malmö now operates nine Avro RJ 100s and three BAe 146-200s.

The Top American Operator – Northwest Mesaba
On 29 October 1996, BAe Regional Aircraft announced a firm $300 million order for twelve Avro RJ85 jets with an option to acquire up to a further twenty-four, from Minneapolis-based Northwest Airlines. These aircraft would be configured in a sixty-nine-seat, two-class layout and operated by Northwest Airlink partner Mesaba Airlines. Mesaba is the fifth largest regional airline in the United States and provides service to 103 US and Canadian cities from Northwest Airlines' three major hubs in Minneapolis/St Paul, Detroit and Memphis. Deliveries commenced in April 1997 with N501XJ (E2208), the former G-ISEE (RJ85 development aircraft and last Hatfield-built aircraft), and continued at a rate of approximately one aircraft per month until the delivery

The RJ in Service

of N512XJ in April the following year. On 30 January 1998 there was a special handover ceremony at British Aerospace Woodford to mark the delivery of the 100th Avro RJ – the ninth Northwest Mesaba aircraft.

Following its earlier success with Air Wisconsin (i.e. a carrier offering regional services for a national airline in a unionised environment), BAe were justifiably pleased to receive a large order from another US regional. This was especially encouraging to the manufacturer after the financial problems experienced by Business Express, which only received three of its planned order for twenty RJ70s before it had to restructure and cancel the order. These three returned aircraft now fly with Air Baltic.

On 21 July 1997, Northwest Airlines exercised their options for twenty-four more RJ85s, bringing the total number of aircraft ordered to thirty-six, making it the largest Avro RJ customer. Deliveries of these additional aircraft began in May 1998 and continued until May 2000.

Mesaba and Scope Clauses

As the deregulation hit in the USA, the larger carriers began to take over the smaller regionals and the pilots' union sought to maintain their pay as the regionals generally had non-unionised, less well-paid crews.

So, the union agreed scope clauses with the airlines, the effect of which was that aircraft larger than seventy seats are flown with unionised crews but non-unionised, less well-paid crews can operate aircraft with fewer seats. In this constrained environment, the Avro RJ was one of the more successful Regional Aircraft in the 1990s, with the sale of thirty-six aircraft to Northwest Mesaba.

It was this union insistence which resulted in Mesaba's RJ85s having only sixty-nine seats. Mesaba capitalised on this by being the only regional airline in the United States to offer a first-class cabin and a seat pitch that was unrivalled in the Regional Airline industry, offering more leg room than the larger 757-300s flown by unionised crews. Mesaba demonstrated the comfort of this arrangement with a cabin cross-section on its website.

In Service in the Rest of the World

Turkey

Turkey's national airline, Turk Hava Yollari (THY), ordered five RJ100s at the Paris Air Show in June 1993. The first RJ100 was delivered soon afterwards in July 1993, and the others followed at the rate of one per month. THY uses its RJ100s on intensive domestic and international routes from Istanbul and Ankara, averaging over eight hours' utilisation a day. More RJ100s were ordered in 1994 and 1995, bringing the airline's fleet to ten. In October 1995, an order was placed for four RJ70s, which had initially been allotted to Business Express. The airline has had several landing accidents with the RJs and now only has eight RJ100s and three RJ70s in operation.

First RJ Certified in Former Soviet union

Uzbekistan Airways, the flag carrier of Uzbekistan, ordered three Avro RJ85s in late 1996, to replace its Yak-40s on domestic and regional routes serving the historic cities of Bukhara and Samarkand, which are particularly popular with visiting tourists. It received its first RJ85, UK-80002, from BAe Regional Aircraft in July, and UK-80001 and UK-80003 were delivered in December 1997. UK-80001 also functions as a VIP aircraft for the Uzbek government.

Bahrain Defence Force

In a contract valued at $25 million, an order was announced for an Avro RJ85 by the Bahrain Defence Force in September 2001. Previously announced as being for an undisclosed customer, the RJ85 was selected for its outstanding airfield performance and operational flexibility. It is utilised for multiple roles, including various transport tasks. Registered as A9C-BDF, the Avro RJ85 was the first civil fixed-wing aircraft to be deployed by the Bahrain Defence Force and was delivered in November 2001. The Bahrain Defence Force increased its fleet to two aircraft in late 2004 with the purchase of one of Azzura Air's RJ85s, which had been stored at Exeter in 2003 and is now registered A9C-HWR.

Uzbekistan Airways RJ85 UK-80002 (E2309) with pre-delivery registration G-6-309, which was delivered on 7 July 1997. Behind it is one of Mesaba's fleet of RJ85s. (Derek Ferguson)

15

Accidents and Incidents

Though the BAe 146/RJ has been in service for more than twenty years, it has an exemplary accident record, with only eight accidents where the aircraft has been written off. Of these, six have resulted in fatalities.

Four of the accidents were caused by runway overruns, three were the result of badly executed approaches where the aircraft hit the ground, and one most unusually was caused by the shooting of both the pilots by a deranged former airline employee.

There have been a number of major incidents, notably the accident to Queen's Flight 146 at Islay in 1994, and the emergency landing at Stansted by a No.32 (The Royal) Squadron machine after a serious engine maintenance error (see Chapter 8). There was also the skilful crash landing of the Air Botswana aircraft at Harare after it was the victim of a windshear.

The unusual and tricky 'Rollback' engine problem during certain icing conditions could have caused serious incidents, but after diligent work by the manufacturer this danger was eliminated.

Though not unique to the BAe 146, there has also been the very occasional incidence of fumes entering the cabin, which caused concern but has now been rectified.

BAe 146 Ingests Man at Hatfield

Only one major accident occurred to the type prior to delivery. That was the very unpleasant occurrence at Hatfield in August 1988 when an aircraft fitter was sucked into an engine of a 146-300, temporarily registered G-BOWW (E3120). He was killed during engine runs. One former employee commented, 'I got to the apron about five minutes after it happened to see a tarpaulin laid over the remains just in front of the engine, Fire department and paramedics in attendance, some people being led away – clearly very distressed. Stunned silence everywhere. I quietly went back to work. Rumour was rife, as you can imagine. Apparently the chap had been doing some sort of pressure test under the engine. The engine was at full power… he just stepped out from underneath, towards the front and was picked up by the airflow'.

The deceased had apparently been wearing some overalls, which had acted as a sail and contributed to the accident. A number of former BAe staff commented on how, prior to this accident, employees had been accustomed to walking around the aircraft under engines while they were running. Thereafter large yellow metal guards were placed in front of the powerplants during engine runs. (The engines are so quiet they do not need silencers!)

Murder in the Air

The first accident to befall the BAe 146 in airline service was the stuff that films are made of, and perhaps appropriately it happened on the west coast of the USA. On 7 December 1987, a PSA employee, David Burke, who had been dismissed for stealing money, used his airport clearance to gain access without going through a security checkpoint and went to N350PS, on which his former supervisor, Ray Thomson, was flying. When the airliner reached cruising altitude, Burke dropped a note into the supervisor's lap stating, 'It's kind of ironical, isn't it? I asked for leniency for my family, remember? Well, I got none, and now

When a disgruntled former employee shot the pilots of PSA's N350PS (E2027) on 7 December 1984, it dived into the ground, killing all on board. Here is the aircraft in happier days in California. (Author's collection)

One of LAN-Chile's fleet of three 146-200s. On 20 February 1991, CC-CET (E2061) overran the runway at Puerto Williams Airport on Tierra del Fuego, and was almost totally submerged in the Beagle Channel. The remaining two aircraft now serve on the British register with Flybe. (Author's collection)

you'll get none.' Burke shot him with a Magnum 44 and then proceeded to the cockpit and shot both pilots. The aircraft went into a steep dive, reaching a velocity of Mach 0.85 and crashed near Paso Robles, California, killing all thirty-eight passengers and five crew.

Runway Overruns

LAN-Chile, which operated BAe 146-200s for seven years between 1990 and 1997, lost one of its three aircraft in February 1991 at Puerto Williams Airport on Chilean Tierra del Fuego. CC-CET overran the runway, ending up in the Beagle Channel (named after Darwin's ship, HMS *Beagle*) that separates Ushaia in Argentinean Tierra del Fuego from Puerto Williams. Twenty of the seventy-three on board died.

Two years later, in July 1993, China North West's 146-300 B-2716 made an inadvertent flapless take-off at Yinchuan Airport and crashed, causing fifty-five fatalities amongst the 113 occupants. The aircraft did not rotate at Vr speed and when the pilot desperately raised the nose, this only caused the tail to drag along the runway. The aircraft did not become airborne and collided with some earth banks before ending up in shallow water.

A Chilean newspaper photograph of CC-CET, almost totally submerged in the Beagle Channel on the day after the event. (Ken Pye)

In further examples of the same kind, Turkish Airlines (THY) lost two of their BAe RJs through landing accidents, though, fortunately, in each case there were no fatalities. In January 1988, TC-THF overran a runway at Samsun and continued down a slope for over 200ft, and in April 2000, TC-THL overran a wet runway on landing at Siirt. The first aircraft was bought by BAe for spares and shipped back to Woodford, while the wreckage of the second is now stored in Istanbul.

Accidents on Approach

On 25 September 1988, EC-GEO of Pauknair, which was a very early production 146-100 and had previously flown with Dan-Air, took off from Malaga for a flight across the Mediterranean to the Spanish enclave of Melilla on the Moroccan coast. The plane crashed into a hill at an altitude of about 2,000ft, six miles from its destination, killing all on board. Pauknair decided not to resume operations and their second aircraft EC-GEP now flies with BA CitiExpress.

Accidents and Incidents

Crossair Flight 3597 departed Berlin-Tegel on the evening of 24 November 2001 for a flight to Zürich. On approach to Zurich-Kloten, the Avro RJ85 HB-IXM descended below the glidepath and a little later the radio altimeter reported 500ft followed by a 'minimum' warning. The captain then ordered a go around, but this was too late. The aircraft struck trees and crashed. Twenty-four of the thirty-three passengers and crew were killed.

THY lost yet another Avro RJ in January 2003 when TC-THG crashed while on approach to Runway 34 at Diyarbakir in limited visibility conditions due to fog. The aircraft broke up and burst into flames after hitting the ground some 120ft short of runway 34, which was not equipped with ILS. Only five of the eighty passengers and crew survived.

Taxiing

Another write-off was Avro RJ85 N528XJ, which suffered severe damage during a taxiing accident at Memphis in October 2002 when it collided with the terminal after maintenance. The number one engine was torn off, and the nose section was bent to the right. Initially Mesaba and BAe discussed the repair of the aircraft with the nose from uncompleted RJX85, but this was abandoned and N528XJ was scrapped in September 2004.

Incidents

Besides the write-offs of the aircraft, there have also been incidents that were potentially serious but which fortunately were not.

Windshear

Air Botswana received a single BAe 146-100 A2-ABD on 9 November 1989. On 12 March 1998, while on lease to Air Zimbabwe, it was involved in a serious incident. Captain Ray Sherwood and F/O Craig Rable were operating a flight from Victoria Falls to Hwange. Approximately 21nm from the airfield in heavy rain, the aircraft was suddenly struck by a massive windshear and descended very rapidly towards tree-top level. The crew just managed to keep the aircraft flying despite stick shaker warnings and climbed away, striking some tree tops in the process.

But damage had occurred, leading to a double hydraulic failure. The captain decided to divert to Harare, as Hwange was a remote airfield and the nearest medical facilities were 75nm distant. At Harare, observers on the ground reported that the main gear was down but the nose gear was up. So, the crew concentrated on getting the right main gear to lock down and the aircraft carried out a number of violent manoeuvres to see if they could achieve this. But by now it was almost dark and fuel was low so the 146 had to land. Meanwhile, the Harare runway was foamed for a nose-gear-up landing.

Air Botswana's A2-ABD (E1101) made an emergency landing at Harare Airport on 9 November 1989 after being struck by a violent windshear and forced into trees at Hwange, which damaged the aircraft's hydraulics. Fortunately, there were only a few minor injuries. (Dan Gurney)

The landing was made on the left mainwheel only and the right mainwheel was only gently lowered onto the ground. The right main gear held and in fact was down and locked. The nose was held up clear of the runway as long as possible and at approximately 80 knots, as elevator control was decreasing, Captain Sherwood gently lowered the nose onto the runway. The ground contact was juddery and noisy but the aircraft was brought safely to a stop. Apart from a few minor injuries the fifty-nine passengers and five crew escaped unscathed.

Following these events, A2-ABD returned to Woodford for repair and is still in service in Botswana.

In a similar incident, Flightline's G-BPNT was struck by a windshear on approach to Florence on 3 June 2004, and the rear fuselage hit the ground. The aircraft made a go-around and landed safely. The aircraft was badly damaged and a new rear bulkhead from an uncompleted Avro RJX had to be fitted before the aircraft could return to service more than six months later.

Control Problems

In December 2002, British European 146-200 G-JEAX was en route from Birmingham to Belfast when the flight crew noticed that the aircraft was oscillating in pitch more than was customary. The captain disengaged the autopilot and was immediately aware of a strong pitch-up tendency. He applied an increasing forward pressure on the control column and supplemented this with nose-down electric elevator trim. Having pitched to below the straight and level attitude, the captain then tried to counter this with a progressive rearward force on the controls. Both pilots then pulled back with considerable force. The control column suddenly moved aft, the aircraft pitched up and the flight crew felt a violent shudder through the whole airframe that lasted for two or three seconds. After this the control forces returned to normal and they were able to level the aircraft. During the pitching manoeuvres, two of the three cabin crew sustained serious injuries. It was later decided that elevator servo-tab icing probably caused these control difficulties.

Double Engine Failure

In an incident on 7 August 2004 Swiss Flight 725 departed Amsterdam on a flight to Zürich. Near Frankfurt, HB-IXU's left in-board engines suffered an uncontained failure and the left outboard engine also had to be shut down when it generated a fire warning after being hit by debris from the failed powerplant. Both engines were shut down and a safe emergency landing was carried out at Frankfurt-Main. None of the fifty-six occupants was injured. The only similar incident was one suffered by a SAM RJ-100 in 1999, when a catastrophic engine failure led to damage to the flaps, wing and engine mounting.

'Rollback'

The term 'rollback' was coined to describe a phenomenon which a few 146 operators encountered in icing conditions close to large thunderstorms at high altitudes and at relatively high air temperatures (and outside the certification envelope). It occurred in conditions of light icing, sometimes not even sufficient to trigger the ice warning or to prompt the crew to select ice protection systems. Its characteristics were a slow reduction in fan speed, lack of response to the throttle, increase in temperature and decay to sub-idle speed. Once detected, the engine would recover itself after descent to below the icing level had been made. However, in some cases, the engine temperature increased towards the limit before this could be done and the engine had to be shut down.

Incidents were usually confined to only one or two engines, but on one occasion, in the Far East, all engines 'rolled back' at about the same time and the pilot elected to shut down all four. Usable power was not regained until 1,500ft above the water. By May 1998 there had been some twelve recorded instances of 'rollback' in over 3 million flying hours, so, although serious, the problem was not widespread. It was, however, judged by the FAA to be unacceptable and an Airworthiness Directive was issued to restrict the maximum allowable operating height of the aircraft to 26,000ft in icing conditions.

Several years' worth of rig and flight testing, including thunderstorm penetration, culminated in proposed modifications to improve ice protection in the engine fan section. A modified engine was installed in G-LUXE (E3001) for what was hoped would be definitive comparative testing. The greatest technical challenge was the accurate measurement of atmospheric data. The probes measuring water content, temperatures, thermo-couples, etc., that were chosen would potentially give a complete picture of the conditions encountered, but the majority had never been flown at 146 cruising heights and speeds. Eleven video cameras were also used, including some inside the engines.

G-LUXE, during icing trials, with emergency 'jettisonable' service doors, cameras, additional probes, the later LF507 engines and displaying BAe, AI(R) and Avro Test logos. (Derek Ferguson)

The aircraft was based at Panama City in October/November 1992 for trials in the anvil regions of well-developed thunderstorm clouds. Setting two engines at higher power reduced the risk of all engines developing 'Rollback'. After three weeks and sixteen uneventful flights, a 'rollback' finally occurred on an unmodified engine, whereas the modified engine on the opposite wing continued to function normally. The fault was induced by very large storms, and whereas most engines could deal with this, the Lycoming was just too small and would choke up – though it only happen in very specialised conditions.

Some additional refinements were made to the 'fix' and BAe then devised an ongoing modification programme to eliminate the problem entirely. Essentially the modifications involved replacing some parts from the ALF502 with those developed for the later LF507 engine, which is fitted on the Avro RJ and on some late-build BAe 146-300s, and which is not subject to the 'rollback' problem. The FAA estimates the modifications will cost US$75,000 per engine.

The aircraft was possibly the best equipped ever for high-altitude flight and the storms penetrated were bigger and more developed than those typically studied during airborne research projects. The trials added considerably to the understanding of cloud physics and, following its experiences in that test role, G-LUXE now serves with the Facility for Airborne Atmospheric Measurements.

16

The RJX and the End of Production

The Short-lived Aero International (Regional)

In January 1995 British Aerospace announced that, together with ATR, it was forming Aero International (Regional) AI(R), a joint aircraft manufacturing and marketing business. ATR is itself a joint venture between Alenia of Italy and Aerospatiale of France, which produces the very successful ATR 42 and ATR 72 twin-turbo-prop feeder liners. British Aerospace took a third share in the new business, which was headquartered in Toulouse. The joint venture was to manufacture the ATR, Avro RJ and the Jetstream 41 families. However, production of the third type in the BAe Regional Aircraft portfolio, the Jetstream 61 (former BAe ATP), was to end.

In January 1996, the company was launched and finished its first business year with a modest result, logging a total of fifty-nine new orders from its three different airliner families, including twenty-one RJs, and achieving a turnover of £800 million.

Plans were announced for a small jet family with two versions, offering fiftyt-eight and seventy seats respectively. The new Airjet seventy-seater was to enter the market in mid-2001 with the smaller version to follow. But, on 24 April 1998, Aerospatiale, Alenia and British Aerospace agreed to dissolve AI(R) as BAe had withdrawn its financial support of £650 million for the development of the new seventy-seater jet. With the RJ attracting orders and, after many years of losses, breaking even for the first time in 1997, Airjet would have just provided additional competition to the RJ. Furthermore, the demise of Fokker was a distinct bonus to BAe as the Fokker 100 had been a major competitor. According to the British Aerospace statement, 'the reason for the disagreement over Airjet was that, on very careful analysis BAe could not see a return on investment from a new regional jet. This view has not changed and is the reason we doubt the viability of all new regional jet programmes.'

The RJX and the End of Production

The attempt to produce a powerful European Regional Aircraft company had come to naught, as it proved impossible to launch a new, common programme for a seventy-seater jet. With the end of AI(R) British Aerospace was free to further develop its own Regional Jet family.

RJX Proposals

During the 1990s, British Aerospace had continued to examine developments of the RJ – re-examining the original twin-engined 146-NRA of 1991 and other ideas. BAe decided to keep the existing four-engined configuration but install a new engine. The engines under consideration for this new aircraft, designated the RJX, were both American; the Allied Signal AS907 and the Pratt & Whitney PW308.

The original idea was that the RJs could be retrofitted with the new engine, freeing up the RJ's LF507 engines that would then be fitted to the 146 fleet. There were some hybrid aircraft in existence like this, such as the 146-300s delivered to China in the early 1990s, which had LF507-1H engines. These were a version of the engine fitted to the RJ but without the FADEC control.

The test programme was planned on fitting a single new engine to G-LUXE (E3001) as a test bed then retrofitting a production RJ with the new engine. However, the programme moved on from this and BAe decided that E3001 was not a suitable testbed and other improvements should be made to the design to make it more attractive to customers.

The Launch of the RJX

On 21 March 2000 BAE Systems 1 formally launched the Avro RJX. The firm had identified that it could re-vamp the aircraft around an improved powerplant from a comparatively low investment (reportedly $100 million or £65 million). The RJ was in need of development as it had been in existence for eight years and new competition in the form of the Canadian Bombardier CRJ 700/CRJ900 and Brazilian Embraer 170/190 were both gathering large orders, though these suffered after 9/11. The 100-seater Boeing 717 (a major re-vamp of a rather elderly McDonnell Douglas design) had entered service the previous year and there was the seventy–eighty-five-seat Fairchild Dornier 728 and eighty-five to 110 seat 928, which appeared to be strong competitors with large orders from established RJ customers, Lufthansa and Crossair, due to enter service in 2003 and 2005 respectively.

For BAE Systems to compete most effectively with the newcomers, a switch to a twin-engine configuration would have been necessary but would apparently cost around $500 million, a sum that BAE was unwilling to spend. As in the early 1990s ,BAE Systems would not finance a major reworking of the design into a twin-engined version as this would require total re-engineering of the wing, a new engine pylon position, increased sweep to increase mach number and changes to the tail. Such a strategy was not consistent with the company's strategy. BAE would invest in major new projects with Airbus – that was the product to invest in, take the risk and get the return. There was to be no major investment in the firm's own airliners but it could carry out limited improvements to existing product while these were still viable. The basic 146/RJ/RJX design was scheduled to continue in production until at least the second half of the decade.

The RJX was a low-cost, low-risk, high-gain development approach. 'Costing $1.5 million more than the earlier RJ, the RJX will have a list price of only about $27 million to $29.5 million – less than for any other comparable aircraft', claimed Nick Godwin, Vice-President Marketing for British Aerospace Regional Aircraft. The RJX was much lighter and cheaper than the Boeing 717 and the Airbus A318, which were both downsized versions of larger, heavier airliners. (On 14 January 2005, Boeing closed its Boeing 717 programme as it had failed to attract many orders.)

Certain commentators criticised BAE Systems for waiting so long to make this decision, which they had apparently been contemplating since 1998, stating this delay had lost the firm many potential orders. In the latter part of the decade orders trailed off – from 1994 to 1999 at least twenty RJs had been delivered per year but by 2000 this had fallen to fourteen and in 2001 only ten were delivered.

The features of the Avro RJX were as follows:
- BAE Systems guaranteed that the RJX would offer customer airlines greater range, lower costs, increased reliability and the ability to operate at higher weights from restricted airports.
- The Honeywell AS977 engine was expected to offer 15 per cent lower fuel burn, 18 per cent better maintenance rates, and lower exhaust noise compared with the LF507 fitted to the RJ. (The Honeywell AS977 engines would fit into nacelles that were was about 50 per cent longer than the existing nacelle for the LF507.)
- BAE Systems retained the option of three different cabin sizes of the 146/RJ in offering seventy to 128 seats, depending on overall length, and five- or six-abreast layout.
- An Avro Business Jet variant was also offered to corporate operators.

Selling the RJX

European carriers and operators in Asia and the South Pacific region were target markets. On 3 April 2000, Druk Air (Royal Bhutan Airlines) became the launch customer, when it ordered a pair of RJX85s to replace its two BAe 146-100s. The new aircraft were to open new routes from the high-altitude airports in the Himalayan kingdom. British Airways franchise operator CityFlyer Express took

The first RJX development aircraft, G-ORJX (E2376), an Avro RJX85 in final assembly at Woodford. (Ian Lowe)

options on six RJX100s in July 2000, to complement its fleet of sixteen Avro RJ100s and 146s.

On 1 March 2001, British European, an established 146 operator, placed an order valued at $600 million for twelve RJX100s and options on a further eight. The UK carrier intended to configure the aircraft for 112 passengers on its own routes, and for 100 travellers on services it flew as an Air France franchisee. First deliveries were to be in April 2002 stretching on until 2006.

RJX Production

The assembly of G-ORJX, the RJX85 prototype (E2376), started at Woodford in March 2000, with engine runs due to be completed a year later. However, the first flight was delayed by the protracted development of the engine/nacelle/pylon package. This 'Integrated Powerplant System' (IPPS) package was developed by BAE Systems in partnership with the AS977 engine supplier Honeywell and GKN Aerospace on the Isle of Wight, but took even longer to produce than a previously revised schedule had anticipated. As a result of this delay, significant early testing of RJX systems was completed in December 2000 with auxiliary power alone and the Avro RJX only commenced engine runs on 18 March 2001.

RJX Maiden Flight

The test plan for the first flight was worked out during early 2001 and this plan flown on the design simulator. In addition, the opportunity was taken to try out any drills following engine failure close to the ground. This was the major worry on the initial flight, especially just after take-off. The result of these tests was that if the engine failure occurred above 1,500ft then a safe return could be attempted.

On Friday, 27 April, high-speed taxi runs were carried out, which consisted of accelerating to high speed then raising the nose. These tests were primarily to allay any fears that the longer and deeper inlet on the AS977 engine might cause disturbed airflow into the engine. Following the successful conclusion of these, the decision was then made to carry out the first flight on the following day.

So, after a five-month delay, the first flight of the Avro RJX85 prototype took place on Saturday, 28 April 2001. G-ORJX lifted off from Woodford Airfield at 12:16 p.m. and remained airborne for two hours and fifty-four minutes. It completed the planned tests and reached a height of 20,000ft and a speed of 250 knots. The flight crew were Avro RJX Test Pilots Alan Foster and Mark Robinson, with Flight Test Engineers Paul Bayley and Derek Ferguson.

After the flight, Alan Foster said, 'Today's first flight went as planned with all tests easily completed, which bodes well for the rest of the programme. The aircraft handled beautifully throughout the flight, displaying the RJX's heritage, complemented by the advantages of the new engine. Any pilot converting from the Avro RJ or BAe 146 will feel immediately at home in the RJX. The most obvious differences from the earlier aircraft were the lower fuel flow rates and the improved climb rate due to the increased thrust at altitude. This is an aircraft we will all enjoy flying over the coming months during the intensive test phase. It is a tribute to the engineering team that created it.'

Nick Godwin, Senior Vice President, Marketing and Communications, said, 'We are confident that the RJX will be at least a year ahead of any similarly-sized competitor, enabling us to build on our excellent customer base of 400 aircraft with over fifty operators. The Avro RJX has already secured fourteen firm orders and fourteen options.'

This prototype was due to undertake the bulk of the performance testing, including a three-month stint in the Americas. Upon return to the UK in late summer, BAE planned to continue airworthiness trials until the aircraft gained JAA certification in the first quarter of 2002.

Public Debut, Duxford

When the test team heard about the RJX Programme Launch at Duxford they treated this with a fair amount of scepticism and humour. Most of this related to

The RJX and the End of Production

G-ORJX, the Avro RJX85 development aircraft, landing after its maiden flight on 28 April 2001. Note the greater length of the AS977 engine nacelles in contrast with the nacelles fitted to the 146s and RJs. (Ian Lowe)

A pleased first-flight crew with G-ORJX. From left to right: Alan Foster (Test pilot), Derek Ferguson and Paul Bayley (Flight Test Engineers) and Mark Robinson (Test Pilot). (Ian Lowe)

flying a brand new aircraft into a museum – which was unfortunately to prove such an apposite prediction.

So the RJX made its public debut (and only truly public appearance) at the Imperial War Museum's airfield at Duxford, Cambridgeshire, on 21 May. G-ORJX flew in from Woodford with British European titles on the right side and performed a demonstration for airlines, suppliers and media who had gathered from around the world for BAE Systems and Honeywell's 'RJX Celebration'. It then landed and parked in front of the Norman Foster-designed American Air Museum, which made an impressive backdrop to the aircraft.

Air Botnia flew in OH-SAH, their newly accepted RJ85, while vintage British-built aircraft offered twenty-minute flights and a flying display was laid on for the guests. At the event, Mike O'Callaghan, Managing Director for BAE Systems Aircraft Services Group, referred to the improvements – that the RJX would offer 17 per cent more range and 10-15 per cent lower fuel burn (somewhat less than the original expectations) combined with 20 to 30 per cent lower engine costs.

The following morning, G-ORJX departed for Woodford. During the short twenty-minute flight the crew noted a fuel imbalance and on taxiing in were told to shut down the No.2 engine quickly. A large fuel leak was found in the oil/fuel heat exchanger, so they were extremely fortunate that the engine had not caught fire.

Testing

By August, BAE Systems Regional Aircraft had completed four months of test flying on the new Avro RJX85. The aircraft had completed final tests to clear its full altitude and speed envelope, and had flown at speeds up to Mach 0.80 and 360 knots to demonstrate a margin over flight manual capabilities. The aircraft had a noticeably better performance than the RJ, especially at high altitude. Between its maiden flight on 28 April and 5 August, the initial RJX85 development aircraft had accumulated almost 130 hours in some sixty test flights.

The first Avro RJX100, G-IRJX (E3378), flew on 23 September 2001. The crew for the flight was Bill Ovel, Pete Lofts, Paul Baylcy, Glen High (Honeywell) and Derek Ferguson. The flight was slightly longer than the first flight of E2376 at three hours and ten minutes. The larger aircraft had been due to take to the air before the end of July, but was kept waiting for improved powerplants. This airframe was also to participate in the performance, handling and systems development trials, and join the RJX85 in autoland validation testing and simulator data collection. The RJX-100 flew into London City Airport on 5 November to demonstrate its compatibility with the 5.5° steep approach on the 3,900ft runway. BAE planned to furnish both prototypes for customer delivery once they had completed testing during the first quarter of 2002.

The RJX and the End of Production

The first Avro RJX100 G-IRJX (E3378) made its maiden flight on 23 September 2001. It is seen here taking off with a nose probe fitted for testing purposes. (Ian Lowe)

The bulk of flight testing was to involve the first two aircraft, including engine handling and FADEC development; auxiliary power unit, performance; flutter, aircraft handling; automatic-landing, simulator data; hot and cold weather and hot-and-high altitude testing; and European Joint Aviation Authorities assessment.

During 2001, the Honeywell AS900 series turbofan accumulated hours rapidly, as flight testing on the RJX and the Bombardier Continental Jet continued for their planned certification in 2002. The engine test programme also involved more than 500 hours of flight time aboard Honeywell Engines Boeing 720 flying testbed N720GT, which had a nacelle mounted on the right side of the front fuselage.

Meanwhile, the first development aircraft, G-ORJX, departed Woodford on 25 August for Williams Gateway Airport in Mesa, Arizona, for hot and high trials and then onto Toluca. During the transit to the USA, a check of the fuel burn revealed that the engine was burning too much fuel – about 2 per cent above the specification. It was still better than the RJ but not as much as it should have been. To the puzzlement of the test crew, they were ordered to return home even though they had not completed the tests, and arrived at Woodford on 22 November 2001.

The End of the RJX Programme

BAE Systems Chief Executive, John Weston, made a surprise announcement on 27 November 2001; 'We have now completed a detailed assessment of the probable impact on our business of the recent severe downturn in the commercial aerospace market. Since September, the trading outlook in these markets has changed substantially. In particular, operating profit expectations for

The Flight-Test crew with G-ORJX, which was based in the USA from August until November 2001, for hot and high trials, first at Williams Gateway Airport in Mesa, Arizona and then in Toluca. (Ian Lowe)

Airbus next year have been reduced significantly and the outlook for Regional Aircraft has deteriorated sharply. Regrettably, it has been concluded that our regional jet business is no longer viable in this environment.' At the various BAE sites, 1,669 employees were to lose their jobs. Closing the RJX programme resulted in a charge of $352 million (£250 million) in 2001 after the firm had spent millions to launch it.

And this is how the staff heard it:

'…we were told to gather in the hangar at 8:30 a.m. Rumours had already started that morning when it had been leaked to the press that BAE SYSTEMS was pulling out of the Regional Aircraft business. Sure enough, we assembled in the darkened hangar like condemned men. The announcement was made that the programme was finished and that we would be sent home for the rest of the day. Didn't feel very much really. The announcement was so out of the blue that it was difficult to come to terms with. The following day we were told that the flying programme would continue in case the company had to fulfil its orders with British European and Druk Air. It sounded to us like the airlines had not been consulted on the impact that this would have on their business. The feeling was that the company had wanted to shut down the business for many years and the events of 11 September 2001 and the subsequent downturn in the airline market gave them the perfect excuse.'

Evidently the decision was made at short notice because in the December edition of BAE's in-house publication, *Response*, (obviously prepared prior to 27 November) the same John Weston declared that the RJX programme remained pivotal to the organisation.

The events of 11 September had knocked potential RJX sales badly – airlines clammed up completely. Even before 11 September, some commentators were of the opinion that the RJX was struggling to gain any significant ground on the rival regional jet families from Brazil's Embraer, Canada's Bombardier and the US-German venture Fairchild Dornier as the RJX only had fourteen firm orders. (On 2 April 2002 Dornier went into administration, leaving the Fairchild Dornier 728/928 in limbo.)

Nick Godwin acknowledged that the RJX had become a victim of a crowded regional-jet marketplace. 'In spite of every effort to drive down costs the price the market currently expects to pay for aircraft of this type (at approximately $30 million each) means we would incur a significant loss on each one sold.' BAE Systems had very successfully streamlined operations in recent years to make the production of even twenty units a year profitable, at least when it had the seventy-ninety-seat market to itself. However, the emergence of new products from Canadair and Embraer, produced in sufficient quantity to allow discounted pricing, increasingly forced BAE Systems to consider selling at below cost price as a means of registering sales.

BAE indicated that it would continue to provide full support for the in-service aircraft, many of which were in the asset management lease portfolio, and complete the four RJs on the line.

Honeywell Aerospace President Bob Johnson said his company's business plan saw a bright future for the RJX, but that BAE Systems was probably unable to realise a return on the aircraft in a reasonable enough timeframe, especially since the worldwide slowdown in airline demand.

BAE's RJX Order Problem

After cancellation, BAE stated that the firm was willing to build all fourteen new RJXs if the launch customers British European and Druk Air wished to proceed with their orders. So testing continued, mostly on G-IRJX. High-risk test flights were not flown unless it was considered necessary. It was felt that risking the crew and aircraft for tests on a programme that might never be finished was not warranted. However, the timeframe of British European's twelve deliveries spread over five years was hardly a viable proposition for BAE in keeping an otherwise redundant production line open. In January 2002 BAE Systems assembled staff for a 'programme update'. They were told that British European and Druk would not be taking the aircraft. Flying stopped immediately and presumably BAE paid the airlines compensation.

The RJX Final Flights

The first production RJX100, which had been destined for delivery to British European in April 2002, made its maiden flight as G-6-391 (E3391) on 9 January 2002. This aircraft was going to participate in the test programme to overcome the five-month delay caused by the delayed gestation period of the IPPS (integrated powerplant system). The following day, all three RJXs took off for an air-to-air photography session, which produced some stunning images.

After that session G-ORJX never flew again, and is currently parked furnished but engineless at the Customer Training School at Woodford. G-IRJX was initially stored and then donated to the Manchester Aviation Heritage exhibition at Manchester Airport. It made its last flight from Woodford to Manchester Airport on 6 February 2003 and is now on public display together with Concorde G-BOAC. G-6-391 had an even shorter life than its compatriots – it flew only five times, the last flight being on 16 January 2002, and was then finally broken up in July 2004. The other seven partly built RJXs were stripped for spares or broken up and the fuselage and parts of one are due to be displayed in the Manchester Museum of Science & Industry during 2005. So the RJX has played its part in the record of the industrial heritage of the area.

The first two RJXs in their element: G-ORJX, RJX85 and G-IRJX, RJX100. The former is in store at Woodford and the latter is on display at the Manchester Airport Museum. (Ian Lowe)

17

Managing and Developing the Assets

Five months after closing the RJX programme, on 26 April 2002, BAE Systems Regional Aircraft announced its launch as a service business. The business would be focused on Customer Support, Engineering and Asset Management. The main business sites were Prestwick (Engineering and Customer Support), Hatfield (Asset Management), Weybridge (Spares) and Woodford (Customer Training and Engineering).

BAE Systems may have ceased to be a civil aircraft manufacturer but the Regional Aircraft division oversees a legacy of more than 1,100 aircraft in service worldwide, over 160 customers and a turnover of £170 million a year. Asset Management has a portfolio of some 450 BAE-owned aircraft made up of 146/RJs, BAe ATPs, Jetstreams and 748s, and as such is the world's largest Regional Aircraft lessor by fleet size and second by value. Its role is to keep those aircraft out on lease as long as possible with the best airlines.

In the early twenty-first century, more than half of the commercial airliners in the world are leased, and leasing companies now purchase the greater majority of aircraft. So, with the cessation of manufacturing, Regional Aircraft still has a powerful role to play in the lease market. Asset Management's role is to ensure that the leased fleet stay in service or are sold, and to broker deals even where these machines are not in BAE Systems' portfolio. More recently, Asset Management has arranged the sale of Boeing 757s and the lease of Airbuses. In mid-2004 Asset Management had seventy 146s on its books with fourteen in store, and thirty-nine RJs of which nine were in store.

BAE Regional Aircraft is also guarding its technical assets and has set up a focus group to help in extending the operational life of the 146 family.

Most recently a life-extension programme was implemented to extend the design life from 55,000–80,000 cycles. The first modified aircraft is now back in service, appropriately with Air Wisconsin, the first airline to order the type in 1981.

As of 30 November 2004 the 146 with the highest number of landings is Air Wisconsin's N606AW (E2033) with 52,490 landings recorded and the highest time aircraft is Ansett's VH-JJP (E2037) with 48,392 hrs and 34,016 cycles

Latest Operators

Airlines function in a dynamic market where mergers, take-overs, bankruptcies and changes of equipment often take place. The whirligig of operators and ownership throws up some interesting combinations, and many 146s have passed through the hands of a large number of operators. The RJ has been much less affected by this and these younger aircraft have endured a smaller than proportional churn in ownership, partly because they were generally sold to more financially sound operators. Even though production has ended, new airlines are becoming operators of the 146/RJ.

In July 2001, Albanian Airlines bought a 146-200 originally delivered to Air Cal, for operation on routes in South Eastern Europe. A larger 300 joined it in February 2003. Both Albanian aircraft are maintained by Hemus Air of Bulgaria, which recently acquired three BAe 146-200s for service in May 2004, previously operated by Ansett in Australia. Hemus Air is an expanding Bulgarian regional carrier operating domestic and international flights, as well as charters. Its 146s will serve a wide range of routes in Central and Eastern Europe, eventually replacing older Russian equipment (predominately smaller Tu-134s and Yak-40s) that have become uneconomic. In a similar part of the world, Romavia of Romania has taken out a lease on a former Meridiana 146-200.

Albanian Airlines 146-200 ZA-MAL (E2054) delivered on 12 July 2001 and photographed at Tirana Airport, together with a Tupolev TU-134. (BAE Systems)

Philippine domestic airline Asian Spirit leased two 146-100s and two 146-200s from BAE Systems in January 2005. They received their first aircraft, RP-C2999 (E1005), in the same month. This aircraft had been flying with National Jet Systems in Australia since 1991, and as G-SCHH was the Far East Tour aircraft in 1984.

Another new operator is Club Air, which commenced operations in November 2002 from its Verona base with two BAe 146-200s. Its prime mission is to develop air transport from Italian regional airports into South Eastern Europe. It currently operates eighty-five weekly flights and carries over 150,000 passengers a year on a rapidly increasing network serving the Balkans. In 2004 it doubled its 146 fleet with the lease of two further Series 200 aircraft from BAE Systems Regional Aircraft. The first of the newly leased aircraft was delivered at the end of May, with the second following in June.

In early 2005 a new Philippines airline Asian Spirit received two 146-100s and is due to receive two 146-200s in 2006; all of these aircraft were previously in store in Australia. In the UK two former Azzura Air RJ70s were delivered to a new airline, EuroManx, which has taken over Flybe's Isle of Man to London City Airport route.

New Venture - Fighting Forest Fires?

Tests took place in September 2004 to see if the 146 could take on the demanding role of air tanker for the US Department of Agriculture Forest Service. A consortium made up of BAE Systems Regional Aircraft, Tronos Canada and Minden Air Corporation, an established air tanker operator, organised a series of demonstrations

An artist's impression of the BAe 146 in a fire-fighting role. Tests took place in the USA in early 2005 with a former China Airlines 146-100 to see if the 146 could be one of the types to replace the Forest Service's ageing fleet of air tankers, which are in dire need of replacement. (BAE Systems)

by a 146-100 in September 2004 to the US fire-fighting community to show its suitability for the very demanding operation as an air tanker.

After a series of accidents in recent years, the Forest Service grounded its fleet of ageing air tankers working on forest fire-fighting contracts across the USA. Some are returning to service but the plan is to replace the fleet – many built in the late 1950s – with thirty-five jet aircraft made up of at least two different aircraft types by 2008.

The flight trials were conducted in Nevada using a former Air China 146-100 N81HN that had been in store at Southend. The instrumented aircraft, ballasted to simulate the 3,000-gallon load of fire retardant, was flown on typical fire-fighting flight profiles to demonstrate its performance and handling in this environment. Minden Air was delighted with the aircraft's performance, particularly its steep descent whilst maintaining low speed, high transit speed, fuel economy and rapid turn round.

A twenty-year-old former Air Wisconsin 146-200, N606AW, was delivered in late January 2005 to Minden Air for conversion to the fire-fighting role. It will have a tank for fire retardant fitted in the cabin and the retardant will be dropped through doors installed in the cargo bay.

From First to Last

Though OH-SAP was the final British airliner delivery, the last civil aircraft delivery from BAe Woodford was on 10 May 2004, when G-LUXE, reconfigured as an Atmospheric Research Aircraft (ARA), was handed over to Facility for Airborne Atmospheric Measurements (FAAM). Previously, the well-known Lockheed Hercules W.2 XV208 (nicknamed 'Snoopy' owing to its extremely long nose probe) had fulfilled a similar role for the National Meteorological Service until withdrawal from service in 2001.

As the original 146 prototype, G-LUXE (then G-SSSH) made its first flight on 3 September 1981 and was converted into the 146-300 prototype taking to the air on 1 May 1987. It continued in this guise until 6 June 2000, when it was put into store having flown 2,915 hours. In December 2000, a contract was signed with the University of Manchester Institute of Science & Technology to convert it into an Atmospheric Research Aircraft (ARA) on a ten-year lease from BAE Systems.

Former 146-100 prototype making its third 'first flight' on 1 October 2003 as the Atmospheric Research Aircraft at Woodford. It first flew as G-SSSH in 1981 – then in 1987 as the 146-300 prototype as G-LUXE. (Ian Lowe)

Work began in July 2001 and, though scheduled for completion in mid-2002, conversion took over two years as it was decided to zero-life the aircraft, which entailed a complete check of the airframe and all equipment. As it was the first aircraft built, its wing was not representative of production models and establishing records on all the equipment was a time-consuming process.

G-LUXE had previously flown under Class B conditions as a flight-test aircraft, but with the decision to have full transport category certification BAE Systems decided to complete the aircraft as near as possible to the standard of China Eastern's 146-300s. These were delivered with 7,000lb thrust LF507-1H engines (similar to the engines fitted to the RJs) to offer better performance.

The aircraft is cleared to 35,000ft but will also spend time as low as 100ft over the sea and as a precaution local strengthening was incorporated in the airframe in the event that the aircraft has to ditch. Amongst the alterations to G-LUXE are the installation of fuel tanks in the wing roots and the forward part of the rear cargo bay to increase the range to 1,600-1,800 miles.

External differences to G-LUXE are the five sensors on two outboard wing pylons, a large pod on the front left forward fuselage and four external TV cameras. Unlike the Hercules that previously fulfilled this role, it does not have a large nose probe but has a series of sensors positioned around the nose, which provide the same measurements.

Internally, the aircraft is full of test equipment and has positions for nineteen scientific crew. Role change may entail the replacement of instrumentation on some of the racks between missions so there will be periods of 'down time' for the installation, calibration and certification of equipment appropriate for a new task.

It was rolled out in August 2003 and made a one-hour-forty-three minute first flight unpainted on 1 October, piloted by Alan Foster and Pete Lofts with Flight Test Engineers Colin Darvill and Tim Bartup. It then embarked on the test programme, which included testing and calibration of the new meteorological equipment. The drag of all this external specialist equipment was 10 per cent greater than calculated so the test programme was extended from eighteen to fifty-seven flights in order to fit and test drag-reduction modifications.

G-LUXE was handed over and delivered to its new base in Cranfield, Bedfordshire, on 10 May 2004 where it is operated by Direct Flight on behalf of the Facility for Airborne Atmospheric Measurements (FAAM). FAAM is run jointly by the Natural Environment Research Council (NERC) Centres for Atmospheric Science and the Met Office.

It is expected to fly in the order of 500 hours per year and may be used by other weather research organisations if time permits. The ARA flew its first overseas deployment to the Azores in July 2004 as part of the Intercontinental Transport

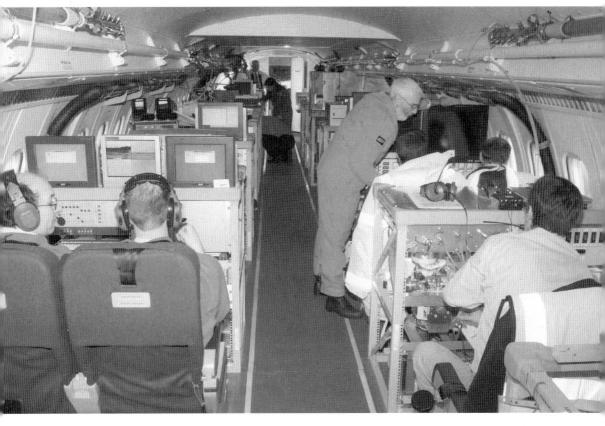

The inside of G-LUXE as an Atmospheric Research Aircraft with positions for nineteen scientists. (BAE Systems)

of Pollutants (ITOP) programme. It centred on using G-LUXE, operating in conjunction with a NASA Douglas DC-8 and National Oceanic & Atmospheric Administration (NOAA) P-3. With the help of sophisticated UK tracking models, US scientists began by making measurements in a polluted air mass off the east coast of the USA. UK scientists then flew out from the Azores to intercept the same air mass and make similar measurements from the BAe 146. During the Azores measurement phase mid-air, wing-tip-to-wing-tip calibrations between aircraft were also carried out during the mission. To complete the picture, German scientists intercepted the air mass in their Falcon aircraft as it landed over Europe. The ARA now has a research schedule taking it well into 2006 and some bookings for 2007.

So the original 146-100 prototype has been through two metamorphoses. The first in 1986-87, when it was lengthened to become the 146-300 prototype, and then secondly in 2001-03, when it was converted for its new career as the Atmospheric Research Aircraft.

18

Conclusion

The BAe 146/RJ: A Summing Up
The BAe 146/RJ proved to be the most successful British-built jet airliner with its 390 deliveries made between 1983 and 2004. What were the critical factors for its success and what held it back?

The Aircraft's Development
At the beginning of the project in the 1970s, the expectation was that the bulk of the sales would be for the short-fuselage 146-100 operating out of small airfields with limited facilities. In 1981, when the 146 first flew, the company was still expecting to sell 225 series 100s and 135 series 200s by 1990. In fact, at the end of 1990 just over 180 of all versions had been assembled. So these forecasts were not borne out, for not only were fewer sold than predicted but also the market sought out the larger 146-200 and 146-300.

However, where people are keen to criticise British industry for being slow to adapt, this criticism cannot be laid at the door of British Aerospace on this occasion. BAe reacted to market demand and speeded up development of the 146-200 so that it was certified only four months after the base model. And in December 1988 the stretched 146-300 entered service. Likewise in 1992, improved versions of the 146 were produced in the form of the re-engined Avro RJ, which sold well.

Perhaps if the RJX had been started sooner it might have achieved more orders. More significantly, if the original twin-engined 146-NRA of 1991 had progressed that might have been a real success – but the financial difficulties of the early 1990s put paid to that.

The Aircraft and the Engine
Though the aircraft proved to be a technical success, in its early years it was let down by the higher-than-average removal rates of the engine. These difficulties

were sorted out, but the memory of these problems stayed with the aircraft and it was only in later years that this impression was finally eradicated.

The aircraft's designer, Bob Grigg, said that he had qualms about whether airlines would buy a four-engined, short-haul jet and the number of engines did prove an impediment to sales, as some customers could not countenance a short-haul airliner with four engines. In an age when medium-to-long-haul jets have only two engines and engine reliability is high, such a view is not surprising – though it was far less apposite when the 146 originated in the early 1970s.

But the engine's quietness also led to many sales just as its superlative airfield performance was a critical factor in the growth of London City Airport and other severely restricted airports.

The Customers

In the early years, the 146 had the misfortune to be adopted by a number of less than financially sound operators, and when many of these went out of business the aircraft were returned to British Aerospace or other owners, which caused bad publicity.

In the United States, though the 146 made quite an impact, deregulation also brought a considerable churn in operators and this instability and the scope clauses which were a by-product of the strongly-unionised workforce militated against the aircraft's success.

The later RJ development generally won orders from better-established, 'blue chip' airlines operating in more stable markets, and as a result many of these are still flying with their initial operators.

Asset Management

In the 1980s, too many 'sales' were actually leases and, when there was a downturn in the airline market, this created a crisis that almost brought BAe to its knees. (In late 2004, seventy BAe 146s and thirty-nine Avro RJs remained on the BAE Systems Asset Management portfolio.) The necessity to manage this challenge led to the setting up of the Asset Management Organisation with its sophisticated financing and marketing skills. Concomitant with this, British Aerospace decided to strictly limit production to approximately twenty units per annum, restricting supply, managing production costs strictly and clearing its backlog of unsold aircraft. Had British Aerospace not resorted to these actions, it might have gone out of business just like its competitor Fokker.

The Competition

The 146/RJ sold more than the rival Fokker 70/100, but the smaller, newer Embraer and Bombardier RJs series have made large sales and have gradually been developed up to 100-seat capacity. Embraer now vies with Bombardier for leadership of the regional jet market. In contrast, the downsized versions of larger airliners including the A318 and Boeing 717 (a re-vamp of the McDonnell Douglas MD-80 and before that the DC-9) have not proved successful in this market.

Following the demise of the Boeing 717 in January 2005, *Flight International* questioned whether the 100-seater market could sustain a number of competing types. In the preceding years the Fokker 100, Avro RJX and Fairchild Dornier 728 programmes had also ended. The only players now left in the market are the slow-selling Airbus A318 and the new Embraer 170/190 series (and Bombardier, which may join with its projected 'C' series).

The End of Civil Aircraft Final Assembly in Britain

In 1993, British Aerospace sold the Corporate Jets Division with its highly successful (Hatfield-designed) BAe 125 to the Raytheon for $372 million and in 1997 BAe closed production of the Jetstream twin-turbo-prop regional airliner. In the late 1990s BAe's venture with the French-Italian Avions de Transport Regional (the short-lived Aero International Regional) fell apart. This might have led to a new joint Regional Jet. But after leaving this grouping, the Avro RJ/RJX had to fend for itself in an increasingly competitive Regional Aircraft market as newer competitors were amassing orders.

The closure of the RJX programme in 2001 marked the end of commercial aircraft final assembly in the UK. BAE Systems decided that its 20 per cent holding in the Airbus was the sole place for its civil airliner investment and this continues to place the UK at the heart of a 'hi-tech' programme which has successfully challenged Boeing's dominance of the world airliner market.

Many British aviation engineers and staff were involved in the programme to design, build, assemble and test the British Aerospace 146/RJ and the aircraft remains a fine tribute to their work.

A large number of the Woodford staff gathered in front of the last RJ (E2394), then registered as G-CBMH, which was later delivered to Blue One as OH-SAP. (See picture in Chapter One). (Ian Lowe)

Appendices

Appendix 1
Flying the BAe 146-300

Our flight was the eleventh for airframe E3120, which is temporarily registered G-BOWW. Divisional chief test pilot Peter Sedgwick kept a keen eye in the right seat. Weight was 74,000lb; rotation and safety speeds were V1 108 knots and V2 116kt, using the mid-take-off flap setting of 24°. Take-off can be made at 30° flap, yet landing setting is only 33°. Trim was mid-to-forward at 31.35 per cent.

All four engines were running in a minute and a half from cold, commendable even for many a twin, and no TGT came within 200°C of the maximum start figure. The pilots' stop-clocks, low outboard on the instrument panels, are fiddly. Ground idle fuel flow, with 25 per cent and 50 per cent N1 and N2, totalled 1,400lb/hr. The fuel cut-off triggers are close-set at the base of throttles, but a simple lever-grip technique avoids any problem of moving two together.

The Garrett GTCP 36-100M auxiliary power unit, used for starting electrical power, fed air to one air-conditioning pack and maintained a good cockpit environment. It has been a standard fit for more than a year on later 200 series aircraft. APU air supply is used on take-off and approach.

The minimum turn radius at the nose-wheel, a little over 40ft, was shown in full-circle swings on the stub of a disused runway. The tail is now the dominant geometry instead of a wing tip, and 16ft clearance must be allowed ahead of the nose for an about-turn from a facing obstruction. The taxiway at the Hatfield base of British Aerospace, which is forty years old and un-resurfaced, provides a testing surface for ground ride. The nose-wheel response, sturdy and sounding like a train over rail joints, was even. Idle thrust is high, and needs regular braking at most weights. At the runway the pitch gust lock on the front of the control column was sprung forward. If forgotten it pulls out of its clip with a gentle tug at rotation.

At maximum thrust the engine fans, though small, have the low whine of bigger units. Take-off on runway 06, with a northerly wind at 15 knots, was easily kept to centreline using the tiller, which has a lot of authority and needed only to be squeezed. The change to controls came at 80 knots, and the ground run took a little over 15 seconds. Rotation is said to be light. I rated it comfortable, and a 10° initial attitude was easily kept in light gusts.

Departure at 2,400ft altitude crossed Stansted airspace under the London TMA. At 250 knots the ride was within the 'cup of coffee test'. A sudden pitch input produced a slight tuning-fork effect more typical of a very long aircraft; the 300 series is more staid in pitch than the 200 and, in particular, the 100.

The dual yaw dampers were selected off – this is clearly warned. A heavy wallow in Dutch roll, kicked in with rudder, could be arrested immediately by the ailerons. The 300 is cleared from previous airspeed restrictions and a 25,000ft altitude limitation, with unserviceable yaw dampers.

Lower cloud tops were 4,000ft. Altostratus from 12-14,000ft, lying between England and a summer sky, held enough moisture to raise an ice warning, and engine anti-icing was selected. The warning master switch is now guarded on; it proved possible to put it off when de-selecting airframe anti-icing. Overall, the master warning panels and aural alerts are well presented.

A climb to the certificated ceiling of 31,000ft was made over East Anglia under Eastern Radar. Airspeed below 10,000ft was 250 knots, which is also the long-range climb speed, and we then accelerated to 280 knots (Vmo is 295 knots). The ceiling is 30,000ft on earlier 146s. On modified 200s and all 300s, the pressure differential above 29,500ft is increased from 6.55 to 6.75 lb/in2, as on long sectors one more flight level (i.e. allowing operation at 31,000 feet) can be an assistance.

At 250 knots, in trim, a pull-up to decrease airspeed by 10 knots and release revealed the long period of the phugoid. The electric pitch trim is powerful but not rapid; 20 per cent offset was held comfortably, and this is not untypical. The trim switches do not respond to a short 'blip'; they must be definitely pushed for a moment, or the manual wheel used. An initially quicker response could feel better.

Control on rudder and manual pitch trim alone is a good test of the delicacy of roll/yaw coupling. They could be comfortably co-ordinated in 15° banked turns, which were readily initiated or stopped with a quick jab of rudder, or adjusted with smoother rudder input.

The control-wheel angle for the break-out of roll spoiler is increased by a degree over the previous 5° setting of the 100 and 200. Most turns are made on ailerons alone, when leisurely banking makes things easy on the passengers. The roll felt perhaps a little stiff from the newness of the airframe. Full wheel response was still in the 146 mould; an evasive manoeuvre could be made very rapidly. Roll spoiler action comes in sharply, with some discontinuity.

Spiral stability is slightly positive to over 35° of bank, and only modest pitch trim is needed for normal turns. The excellent 30° down visibility is a boon in turns, and the 40° upward vision enabled a descending Boeing Awacs to be seen well above us to the side.

The engine is flat rated to 15° at sea level. LP spool speed (N1), rather than turbine gas temperature (TGT), is the limiting parameter above 10,000ft. Thrust was easily set on the gauges; the instrument diameter is not large, but the digits are clear and the display is well damped. The adjacent N1 and TGT digital displays seem similar, and I mistook one for the other in power setting. I was told that everyone does it once.

The thrust modulation system (TMS) can be described as a limited-authority auto-throttle, but it is much more. The synchronisation function allows the pilot to use the handful of four throttles as if they were a single lever. With level angles aligned, the TMS fine-tunes the required parameter or provides synchronisation. It calculates N1 for go-around, or trims to take-off, maximum continuous, or TGT settings on each engine. In synchronisation engine, 1 or 2 can be set as the master for a choice of common N1, N2, or TGT, once setting is close to the selected value.

The cabin is quite altered by the small special changes. The more intrusive lockers still leave aisle headroom over an adequate width, and easier-to-use locker doors reveal deeper bins with improved indirect internal lighting. Most evident in the cabin was the lack of noise from airflow near the flap track. The deep flap-end channels in the fuselage, now faired by sliding plates, have disposed of that 'organ-lip' noise. The noise of initial flap extension remains, and correction is still in hand.

Door-seal noise has been reduced, and the wing/body fairing improved. The AC hydraulic pump is now automatic on demand, and the passengers less often hear its characteristic noise. Engine noise is not intrusive, except for a slight growl around rows 5 and 6 in the plane of the geared fans.

Cockpit cruise airflow makes a characteristic searing sound, which is a little louder at the increased Mmo of Mach 0.70 on the earlier models to 0.72 of the 300. At Mach 0.76, a 45° banked turn was made with a couple of 'g-pulls'. The merest wing burble or aileron complaint was felt in roll at this speed. At the low-speed end, at 70,000lb, the stickshaker operated as expected at 144 knots. The aircraft was still very manoeuvrable 10 knots above this, with only a suggestion of wing airflow becoming irregular at 150 knots.

An emergency descent from 28,000ft to 15,000ft was made under autopilot control. With the throttles closed in a turn, the nose eased down with speed-lock selected. The clamshell speed-brakes were fully selected, but they blow back above 210kt. The speed-brakes hardly induce buffet at high speed, and their extension is prominently captioned. The speed-brakes cannot operate at high engine power. They retract automatically if N1 goes above 70 per cent on approach, and are inhibited above 88 per cent in the cruise. If selected first for emergency descent, they stay stowed until throttle closure starts. To increase airspeed the nose was pushed further down, with the 'synchro' button, to the side of the outboard arm of the control wheel, pushed to reset the autopilot datum. This button is easy to use, and inadvertent operation is unlikely. With the aircraft stabilised in a 7-8° nose-down attitude at 295kt, the descent rate was off-scale at more than 6,000ft/min. We levelled off comfortably in the final 500ft. A 2¼ minute descent equated to 5,800ft/min average.

Stalls were made clean, with 18° flap in a 25° banked turn, and in the landing configuration. The level of wing buffet, with the mid-c.g., made the stickshaker warning secondary. A wing was inclined to go down at the stall, but there was still plenty of roll-spoiler control. Without leading-edge devices, the clean stalling speed is about 40kt higher than when fully flapped. At 68,000lb, the stalling speed in final approach configuration is a mere 88kt, and Vref 108kt, but final flap retraction speed on go-around would be about 160kt. A pneumatic stickpusher is fitted and its action becomes progressively more forceful as flap is extended, but it quickly ceases with airspeed build-up.

Returning to Hatfield, the captain's controls were disconnected from the co-pilot's – as with a control jam. Loads with half control felt little different (the left control retains Q-feel and may appear heavier; g weight compensation makes right control alone lighter). Aileron control was not split, as it can be reconnected only on the ground.

The certificated maximum landing crosswind is 35kt. In the first landing, made with a 15-17kt crosswind component, the docility of the 146 was impressive. Most new trainees could probably handle a 20kt crosswind straight off. The next landing was made in the reverse direction, with opposite crosswind, and the circuit speed with 18° flap was 160kt. Again approaching without speed-brake, 45-50 per cent N1 only was required on finals. The 146 is a bit 'slippery', and one must use flight idle for a time to reduce speed quickly. Approaches are normally made with the speed-brake extended right to landing. Number four engine was pulled to idle on rotation from the touch and go. It was difficult to notice that anything had happened. The minimum control speed is less than 100kt. Circuits with one engine out are a non-event, and normal procedures are used. A circuit was then made with two engines on one side idling. Approach flap is limited to 24° instead of the full 33°, and Vref increased by 11 knots. Untrimmed rudder loads were still remarkably light. Even go-around from 100ft at 120 knots, the minimum allowed speed, produced little load against the stop. A slight drift from the runway centreline was quickly stopped with wing-down into the live engines. Handling was comparable with one engine out on a good twin.

The 146 had recently demonstrated steep approaches into London City Airport so we put this to the test. The first steep element started from only 500ft was comfortably under control, save for my failing to achieve Vref+5 knots until a late stage. The 1,000ft/min descent was killed in a far-from-heavy flare, and the aircraft landed itself in its usual gentle fashion one second later.

A flapless approach was made before the final full stop. At 165 knots the altitude was 5° nose-up on a flattish 2.5° approach. The view ahead became more like that in a TriStar, in stark contrast to the nose-down attitude on normal approach. Float in over-flaring is a common sin to avoid when touching down flapless and at high speed. The elevator is much more effective at the high airspeed, and the aircraft has to be 'put on'. The last approach was the party piece: I held a height of 1,300ft until the threshold was about to disappear under the nose, with 33° flap and the airspeed already set. With speed-brake extended and throttles closed down, we went. Thrust was added as airspeed again stabilised. N1 at 55 per cent gave a good margin over the 40 per cent flight idle.

A headwind component of 5 knots came from the persistent 15-20 knots wind. With a descent rate of 1,150ft/min, and a nose-down attitude well over 5°, the view of our target touchdown was clear and impressive. The slight turbulence allowed Vreft+5 knots (110 knots at 65,000lb) to be kept accurately. Few adjustments to path or power were made until the flare.

With a ground speed of 105 knots, the approach path was angled at just over 6°. I had previously demonstrated the same angle of approach at 74,000lb and Vref+5 knots (119 knots) in the simulator, stopping comfortably in just over half of the visual display's 1,200m runway. For certification to a 6° approach slope, trials will be made to 7.5° for the CAA and 8° for the FAA.

A steeper, slower entry to the flare still resulted in the mildest of touchdowns. The throttles were pulled back through the now released ground-idle gate. Ground-idle selection must be made before the speed-brake lever can be palmed further back to release all wing spoilers up to the lift-dump position. Although the brakes were used sparingly, a long taxi remained down the 1,850m runway. Chock-to-chock 2 hours and 15 minutes, with 2 hours in the air, the fuel burn totalled 9,370lb.

The flight deck of BAe 146-100 G-SCHH (E1005) with Chief Test Pilot Peter Sedgwick in the left seat, and John Creswell in the right. (The flight deck of E3120 flown, in Appendix 1, was essentially the same.) (BAE Systems)

When returning from a later air-to-air photography session, Peter Sedgwick demonstrated a minimum-distance stop. For an aircraft with no reverse thrust, the stopping power is notable. The now-standard carbon brakes cool more quickly than steel units. Brake temperature indication is an option, but fans are now a standard fit, operating normally whenever the nose-wheel is extended, which is good for multi-sector operations.

I must reiterate that the 146 is a pilot's aircraft. It wraps itself around you like a glove. It has a big cockpit, a good view, ample stowages and clips, is forgiving at low and high speed, its thrust is easy to handle, and it has the softest of soft under-carriages.

By Harry Hopkins

(An edited extract reprinted with kind permission from *Flight International*, 22 October 1988)

Appendix 2
BAe 146/RJ/RJX Production List

The final production of the BAe 146/RJRJX was:
- 35 series 100s and 12 RJ70s
- 116 series 200s and 87 RJ85s
- 70 series 300s and 71 RJ100s
- Three 3 RJXs were completed and flown.
- The first 146-100 later became a 146-300 and is now a 146-301.

B Class Registrations
Many of the aircraft made their first flights with B Class Registrations. Initially, Hatfield constructed aircraft adopted the former de Havilland B Class marks, i.e. G-5 and the Woodford aircraft had the former Avro B Class marks, G-11. From late 1989 a standardised system was introduced of G-6 plus the last three digits of the construction number wherever the aircraft was assembled.

N.B. Six more RJXs in various stages of construction were scrapped, except for a composite RJX made of sections from E2396/E3397/E2400 which will be exhibited in the Manchester Museum of Science and Industry.

MSN	Type	Current Registration	First flight	Delivery to current operator	Current/last operator	Previous registrations	Built	Notes
E1001	146-100	G-LUXE	03/09/81		BAe/BAE Systems	G-BIAD/G-SSSH/G-5-300	H	Prototype 100
E3001	146-300	01/05/87						Converted 300 prototype
	146-301	01/10/03		10/05/04	FAAM			Converted Atmospheric Research Aircraft
E1002	146-100QT		25/01/82	30/08/95	Nat.Jet Systems	G-SSHH/G-OPSA/G-5-146/N5828B/ G-5-005/N801RW/G-BPNP/ N720BA /G-BSTA/OE-BRL /ZS-NCA	H	FF as STA 02/08/88 FF as QT 05/12/90
E1003	146-100	VH-NJX	02/04/82	27/08/01	Nat.Jet Systems	G-SSCH/G-5-14/N246SS/VH-NJA/EI-CPY	H	
E1004	146-100	VH-NJA	29/08/82	01/09/02	Nat.Jet Systems	G-OBAF/ZD695/G-5-04/G-BRJS/G-5-537 /G-OJET/PK-MTA/G-DEBJ		
E1005	146-100	RP-C2999	19/10/82	01/05	Asian Spirit	G-SCHH/ZD696/VH-NJY	H	
E1006	146-100	G-OFOA	05/05/83	20/06/98	Formula One	G-BKMN/SE-DRH/EI-COF	H	
E1007	146-100	EC-GEO	02/06/83	21/09/95	Pan Air	G-BKHT/EC-969	H	Cr Melilla 25/09/98
E2008	146-200	G-BMYE	01/08/82		BAe	G-WISC/G-5-146/G-WAUS	H	1st 146-200 Bu Filton 05/95
E1009	146-100	VH-NJZ	05/10/83	03/07/91	Nat.Jet Systems	TZ-ADT/G-BRUC/G-6-009	H	
E1010	146-100	G-JEAO	08/11/83	19/09/94	BAE Asset Mgmt	G-BKXZ/PT-LEP/G-5-512/N802RW/ C-GNVX/G-UKPC	H	Std Filton 09/01/03
E1011	146-100	D-AWDL	17/12/83	30/06/98	WDL Aviation	PT-LEQ/N803RW/C-GNVY/G-UKJF	H	
E2012	146-200	I-TERB	25/05/83	09/10/03	Club Air	N601AW/C-FHAV/G-DEFK/EI-DBY	H	
E1013	146-100	VH-NJC	21/02/84	23/06/90	Nat.Jet Systems	N146AP/G-6-013	H	
E2014	146-200	I-TERV	17/09/83	24/10/03	Club Air	N602AW/C-FHAX/G-DEFL/EI-DBZ	H	
E1015	146-100	G-MABR	16/04/84	31/03/02	BA Citiexpress	G-5-01/N461AP/XA-RST /N568BA / EC-971/EC-GEP/G-DEBN	H	
E2016	146-200	G-DEFM	03/05/84	22/10/99	Flightline	N605AW/C-FHAZ	H	Std Southend 04/10/03
E1017	146-100	G-BLRA	03/10/84	29/09/97	BAE Systems	G-5-02/N462AP/CP-2249/N117TR	H	
E2018	146-200	G-TBAE	09/12/83	18/01/03	BAE Systems	G-HWPB/G-JEAR		N603AW/G-OSKI/G-BSRU/G-018
E1019	146-100	B-2701	07/04/86	01/05/88	China North West	G-5-019/G-5-523/G-XIAN	H	Std Lanzhou 08/03
E2020	146-200	G-JEAS	09/02/84	28/02/96	Flybe	G-BSRV/G-OLHB	H	N604AW/C-FEXN/G-OSUN/
E1021	146-100	ZE700	23/11/84	23/04/86	Royal Air Force		H	
E2022	146-200	G-DEBE	16/05/84	19/08/04	Discovery	G-5-02/G-5-507/G-6-021	H	
E2023	146-200	G-CLHD	31/05/84	03/10/03	Flightline	N346PS/N163US	H	
E2024	146-200	EI-CZO	06/07/84	07/03/03	CitJet	N347PS/N165US/G-DEBF	H	
E2025	146-200	G-TBIC	22/10/84	01/12/96	Flightline	N348PS/N166US/G-DEBC/G-CLHA/G-FLTB	H	
E1026	146-100	B-2702	16/05/86	01/05/88	China North West	N349PS/N167US	H	Std Lanzhou 08/03
E2027	146-200	N350PS	24/11/84	07/12/84	PSA	G-5-026	H	Cr Paso Robles 07/12/87

MSN	Type	Current Registration	First flight	Delivery to current operator	Current/last operator	Previous registrations	Built	Notes
E2028	146-200	D-AMAJ	06/12/84	11/07/00	WDL Aviation	N351PS/N171US/G-DEBA/G-BZBA	H	
E1029	146-100	ZE701	12/04/85	09/07/86	Royal Air Force	G-5-03/ZE701/G-6-029	H	
E2030	146-200	EI-PAT	07/03/85	26/03/99	CityJet	N352PS/N172US/G-WLCY	H	
E2031	146-200	EI-CNQ	07/02/85	18/11/96	CityJet	N353PS/N173US/G-OWLD	H	
E1032	146-100	B-2703	10/07/86	01/05/88	China North West	G-5-032	H	Std Lanzhou 08/03
E2033	146-200	N606AW	19/02/85	27/02/85	Air Wisconsin	N354PS/N174US/G-DEBD/G-BZBB	H	
E2034	146-200	VH-NJW	09/05/85	27/05/00	Nat.Jet Systems	G-5-035/B-2704/B-584L/G-BVUW/J8-VBC	H	
E1035	146-100	G-JEAU	27/11/86	28/01/97	Flybe	N355PS/N175US/HB-IXB/G-CLHB	H	
E2036	146-200	G-GNTZ	24/05/85	07/05/02	BA Citiexpress		H	
E2037	146-200	VH-JJP	26/03/85	01/09/93	Ansett Australia		H	Std Melbourne 27/03/02
E2038	146-200	VH-JJQ	07/06/85	01/09/93	Ansett Australia		H	Wfu Brisbane 06/02
E2039	146-200	EI-CMY	03/07/85	28/06/96	CityJet	N356PS/N177US	H	
E2040	146-200	VH-YAF	18/07/85	16/04/00	Nat.Jet Systems	N357PS/N178US/G-DEBG	H	
E2041	146-200	N179US	07/08/85	01/09/85	Air Wisconsin	N358PS	H	
E2042	146-200	N181US	30/08/85	01/09/96	Air Wisconsin	G-BMFM/N359PS	H	
E2043	146-200	N183US	17/09/85	01/09/96	Air Wisconsin	N360PS	H	
E2044	146-200	EI-CMS	02/10/85	26/04/96	CityJet	N361PS/N184US	H	
E2045	146-200	OY-RCA	24/10/85	07/02/00	Atlantic Airways	N362PS/N185US/G-DEBH/G-CLHE	H	
E2046	146-200	EI-CNB	08/11/85	07/08/96	CityJet	N363PS/N187US	H	
E2047	146-200	G-OZRH	21/11/85	19/03/04	CityJet	N364PS/N188US	H	
E2048	146-200	G-FLTA	03/12/85	25/02/98	Flightline	N365PS/N189US	H	
E2049	146-200	N608AW	21/03/86	24/04/86	Air Wisconsin	G-5-002	H	
E2050	146-200	D-AWUE	12/02/86	03/01/99	WDL Aviation	G-5-004/G-5-517/PK-PJP	H	
E2051	146-200	EI-CWB	28/01/86	29/03/01	CityJet	G-5-003/N141AC/N694AA/SE-DRE	H	
E2052	146-200	N607AW	10/12/85	15/01/86	Air Wisconsin	G-5-001	H	
E2053	146-200	EI-CWC	02/03/86	27/04/01	CityJet	G-5-053/N142AC/N695AA/SE-DRC	H	
E2054	146-200	ZA-MAL	25/04/86	12/07/01	Albanian Airlines	G-5-054/N144AC/N696AA/SE-DRG/G-BZWP	H	
E2155	146-200	D-AMGL	01/05/03	01/02/99	WDL Aviation	G-5-055/N145AC/N697A/SE-DRF/G-CBFL	H	
E2156	146-200QT	EC-HDH	20/03/86	01/01/00	TNT/Pan Air	G-5-056/N146FT/N146QT/G-TNTA	H	First QT
E2157	146-200	G-CBAE	08/07/86	01/02/99	BAE Asset M.	G-5-057/N146AC/SE-DRB	H	Std Exeter 12/01
E2158	146-200	EI-CWA	18/07/86	07/11/02	CityJet	G-5-058/G-ECAL/N148AC/N699AA/ CP-2254/SE-DRI/G-ECAL	H	
E2159	146-200	G-JEAW	28/06/86	28/08/97	Flybe	G-5-059/N401XV/CC-CEJ/N759BA	H	
E2160	146-200	EI-DDE	18/08/86	24/11/97	CityJet	G-5-060/N402XV/XA-RMO/CP-2260 D-AZUR/G-CCJC	H	
E2161	146-200	CC-CET	26/08/90	24/02/90	Lan Chile	G-OHAP/G-5-061/N403XV	H	Cr Puerto Williams 20/02/91
E2062	146-200	EI-CSK	03/12/86	08/08/00	CityJet	G-5-062/N406XV/N880DV/N810AS	H	
E1063	146-100	N463AP	30/10/86	03/04/91	Air Wisconsin	G-5-063/N70NA	H	Std Calgary 07/07/04

MSN	Variant	Reg	Date	Date	Operator	Identities	H	Notes
E2064	146-200	G-JEAV	01/11/86	28/06/97	Flybe	G-5-064/N404XV/CC-CEN/N764BA	H	
E2065	146-200	Z-WPD	18/12/86	13/10/87	Air Zimbabwe	G-5-065	H	Wfu Harare 28/04/00
E2066	146-200	I-TERK	12/12/86	12/12/97	Club Air	G-5-066/N405XV/C-FHNX/XA-RTI/N356BA/D-ALOA/G-CCJP	H	
E2067	146-200QT	OO-TAR	13/03/87	01/01/00	TNT Airways	G-5-067/G-TNTB	H	
E1068	146-100	N114M	20/12/86	25/10/97	Moncrief Oil	G-5-068/B-2705/B-585L/J8-VBA/N861MC	H	
E2069	146-200	D-AEWD	21/05/87	26/03/97	Eurowings	G-5-069/G-BNKJ/N407XV/OO-DJC/G-UKLN	H	
E2070	146-200	N609AW	08/06/87	19/06/87	Air Wisconsin	G-5-070/G-BNKK	H	
E1071	146-100	G-JEAT	29/01/87	11/10/96	Flybe	G-5-071/B-2706/G-BVUY/J8-VBB	H	Bu Exeter 08/03
E2072	146-200	LZ-HBA	28/03/87	01/04/04	Hemus Air	G-5-072/G-BNJI/N366PS/N190US/HB-IXC/EI-CTY/VH-NJQ		
E2073	146-200	LZ-HBB	08/04/87	08/04/04	Hemus Air	G-5-073/N367PS/N191US/HB-IXD/G-BVFV/EI-JET/VH-NJU	H	
E2074	146-200	EI-CSL	19/05/87	19/10/00	CityJet	G-5-074/G-BNND/N368PS/N192US/N146SB/HS-TBQ/N881DV/N812AS	H	
E2075	146-200	OY-CRG	23/06/87	23/03/88	Atlantic Airways	G-5-075/N369PS	H	
E1076	146-100	B-632L	14/03/87	01/05/88	A/C Holdings	G-5-076/B-2707	H	
E2077	146-200	D-AEWE	27/07/87	25/04/97	Eurowings	G-5-077/N408XV/G-BRNG/OO-DJD	H	
E2078	146-200QT	I-TNTC	28/07/87	01/01/00	TNT/Mistral Air	G-UKRH	H	
E2079	146-200	G-MIMA	12/08/87	28/04/02	BA CitiExpress	G-5-078/G-BNPJ	H	
E2080	146-200	N290UE	14/08/87	10/12/97	Air Wisconsin	G-5-079/G-CNMF	H	
E1081	146-100	N81HN	19/05/87	07/01/04	A/C Holdings	G-5-080/N290UE/N814AS	H	
E2082	146-200	N610AW	03/09/87	16/09/87	Air Wisconsin	G-5-081/B-2708/B-633L	H	
E1083	146-100	G-CCXY	04/06/87	01/05/88	Avtrade	G-5-082	H	
E2084	146-200	N291UE	24/09/87	01/04/98	Air Wisconsin	G-5-083/B-2709/B-633L	H	
E1085	146-100	ZA-MAK	03/07/87	01/05/88	Albanian Airlines	G-5-084/N815AS	H	
E2086	146-200QT	EC-GQO	30/09/87	21/10/97	TNT/PanAir	G-5-085/B-2710/B-635L/G-CCLN	H	
E2087	146-200	N292UE	08/10/87	01/04/98	Air Wisconsin	G-5-086/G-BNUA/SE-DEI/D-ADEI	H	
E2088	146-200	G-MANS	21/10/87	16/04/02	BA CitiExpress	G-5-087/N816AS	H	
E2089	146-200QT	OO-TAW	10/11/87	17/01/01	TNT Airways	G-5-088/G-CHSR/G-CLHC	H	
E2090	146-200	C-FBAB	27/11/87	22/11/01	Air Canada Jazz	G-BNYC/G-TNTH/EC-281/EC-EPA	H	
E1091	146-100	A6-SHK	21/12/87	20/12/88	Abu Dhabi Gvt.	G-5-090	H	WFU Bournemouth 01/01
E2092	146-200	C-FBAE	16/11/87	22/11/01	Air Canada Jazz	G-5-091/G-BOMA	H	
E2093	146-200	VH-IJS	08/10/87	01/09/93	Ansett Australia	G-5-092	H	Std Perth 09/01
E2094	146-200	OY-CRB	03/12/87	06/04	Atlantic Airways	G-5-093	H	
E1095	146-100	A5-RGD	26/01/88	16/11/88	Druk Air	G-5-094/G-CSJH/SE-DRD	H	
E2096	146-200	C-FBAF	10/02/88	22/11/01	Air Canada Jazz	G-5-095/G-BOEA/G-6-095	H	
E2097	146-200	VH-YAD	18/02/88	15/12/96	Nat.Jet Systems	G-5-096	H	Std Perth 09/01
E2098	146-200	VH-IJT	26/02/88	01/09/93	BAE Asset Mgmt	G-5-097/N461EA/N293UE	H	
E2099	146-200	G-JEAJ	16/03/88	27/03/93	Flybe	G-5-098	H	
E2100	146-200QT	OO-TAU	22/03/88	01/01/00	TNT Airways	G-5-099/G-OLCA	H	
						G-5-100/G-BOHK/G-TNTJ/D-ANTJ/EC-GQP	H	

MSN	Type	Current Registration	First flight	Delivery to current operator	Current/last operator	Previous registrations	Built	Notes
E1101	146-100	A2-ABD	14/01/89	09/11/89	Air Botswana	G-5-101	H	
E2102	146-200QT	EC-ELT	13/04/88	24/10/88	TNT PanAir	G-5-102/G-BOKZ/EC-198	H	
E2103	146-200	G-JEAK	21/04/88	24/03/93	Flybe	G-5-103/G-OLCB	H	
E1104	146-100	VH-NJE	24/01/89	25/06/96	Nat. Jet Systems	G-5-104/HS-TBQ/CP-2247/G-BTXO	H	
E2105	146-200QT	EC-FZE	10/05/88	01/01/00	TNT PanAir	G-BOMI/G-5-105/HA-TAB/G-TNTP / EC-719	H	First 146 built Woodford
E2106	146-200	C-GRNZ	16/05/88	22/11/01	Air Canada Jazz	G-5-106	W	
E2107	146-200	VH-YAE	24/06/88	01/12/96	Southern Austr	G-5-107/N294UE/N462EA	H	Bu Melbourne 01/04
E2108	146-200	EI-CWD	22/06/88	13/06/01	CityJet	G-5-108/N295UE/SE-DRK	H	
E2109	146-200QT	I-MSAA	30/06/88	15/05/00	Mistral	G-BOMJ/SE-DHM/RP-C481/G-TNTD/I-TGPS	H	Std Melbourne 06/02
E2110	146-200	VH-JJW	15/09/88	01/09/93	Ansett Australia	G-5-110	H	Bu Melbourne 01/04
E2111	146-200	C-FBAO	06/07/88	28/12/01	Air Canada Jazz	G-5-111	W	
E2112	146-200QT	EC-HJH	02/08/88	01/01/00	TNT PanAir	G-5-112/G-BOMK/F-GTNU/RPC482	H	First with EFIS
E2113	146-200	VH-JJY	07/10/88	04/05/89	Ansett Australia	G-BOXD	H	
E2114	146-200QT	G-ZAPR	28/10/88	16/01/04	Titan Airwys	G-BOXE/VH-JJZ	H	
E2115	146-200	SE-DRA	05/09/88	01/92	Malmo Aviation	G-11-115/G-5-115/C-GRNY/GBRXT	W	
E2116	146-200	VH-JJU	23/11/88	01/09/93	Ansett Australia	G-11-116/G-5-116/ZK-NZA	H	Std Filton 01/08/03
E2117	146-200QT	EC-FVY	08/12/88	01/01/00	TNT PanAir	G-BPIS/F-GTNT/G-TNTO/EC-615	H	
E3118	146-300	D-AQUA	22/06/88	11/12/97	Eurowings	G-OAJF/G-6-118/HB-IXZ/G-OAJF	H	Bu Townsville 08/01/03
E2119	146-200QC	G-ZAPN	06/01/89	20/09/99	Titan Airways	G-BPBT/ZK-NZC	H	
E3120	146-300	N611AW	30/08/88	16/12/88	Air Wisconsin	G-BOWW/G-5-120/N146UK	W	
E2121	146-200	C-FBAV	03/12/88	12/12/01	Air Canada Jazz	G-11-121	H	
E3122	146-300	N612AW	15/12/88	23/12/88	Air Wisconsin	G-5-122	H	
E3123	146-300	EI-DEV	10/03/94	01/04/00	CityJet	G-5-123/G-UKHP	H	Std Mojave 19/06/04
E1124	146-100	PK-OSP	14/06/89	09/08/03	Airfast Indonesia	G-5-124/G-6-124/ZE702/G-CBXY	H	
E3125	146-300	G-UKSC	03/03/89	01/04/00	BAE Asset Mgmt	G-5-125	H	
E3126	146-300	G-BPNT	08/03/89	01/04/93	Flightline		H	Std Melbourne 27/02/03
E2.27	146-200	VH-JX	03/12/88	01/09/93	Ansett Australia	G-11-127/G-5-127/ZK-NZB	W	
E3.28	146-300	G-JEAM	23/04/89	26/05/93	Flybe	G-11-128/HS-TBK/G-BTJT	W	
E3.29	146-300	G-BTXN	24/05/89	12/09/04	Discovery	G-5-129/HS-TBM/G-BTXN/EI-CTM	H	
E2.30	146-200	C-GRNX	27/04/89	22/11/01	Air Canada Jazz	G-5-130	W	
E3.31	146-300	EI-CLG	25/05/89	08/06/95	Aer Lingus	G-11-131/G-BRAB/HS-TBL	H	
E3.32	146-300	N614AW	14/05/89	23/05/89	Air Wisconsin	G-5-132	H	
E2.33	146-200	C-GRNV	27/05/89	13/12/01	Air Canada Jazz	G-5-133/G-BPUV	H	
E3.34	146-300	D-AWBA	05/07/89	26/03/00	WDL Aviation	G-11-134/G-5-134/ZK-NZF	W	
E3135	146-300	ZK-NZG	21/07/89	08/06/01	Quantas NZ	G-5-135	H	

MSN	Type	Reg	First flight	Into service	Operator	Previous identities	H/W	Notes
E2136	146-200	G-JEAX	30/06/89	27/03/98	Flybe	G-5-136/N882DV/N136TR/C-FHAP/N136JV	H	
E2137	146-300	ZK-NZH	04/08/89	08/06/01	Quantas NZ	G-11-137/G-5-137	W	Std Essendon 21/02/02
E2138	146-200	G-JEAY	29/06/89	29/03/01	Flybe	G-5-138/N883DV/N138TR/C-FHAA / N138JV/SE-DRL	H	
E2139	146-200	C-GRNU	06/09/89	22/11/01	Air Canada Jazz	G-5-139	H	
E2140	146-200	C-GRNT	25/11/89	22/11/01	Air Canada Jazz	G-11-140	W	
E2141	146-300	N615AW	27/08/89	09/09/89	Air Wisconsin	G-5-141	H	
E2142	146-300	EI-DEW	27/10/89	20/05/04	CityJet	G-5-142/G-UKAC	H	
E2143	146-300	ZK-NZI	13/12/89	08/06/01	Mount Cook A/L	G-5-143	H	Std Essendon 25/02/02
E1144	146-100	G-OFOM	27/02/90	13/03/00	Formula One	G-6-144/G-11-144/G-BSLP/N3206T	W	
E3145	146-300	N616AW	12/10/89	03/11/89	Air Wisconsin	G-5-145	H	
E3146	146-300	EI-CLH	08/01/90	02/06/95	Aer Lingus	G-BOJJ/I-ATSC	H	Std Mojave 02/06/04
E3147	146-300	ZK-NZJ	21/10/89	08/06/01	Quantas NZ	G-11-147	W	Std Melbourne 11/03
E2148	146-200QC	G-ZAPK	29/11/89	25/04/96	Titan Airways	G-PRIN/G-6-148/G-BTIA/ZS-NCB	H	
E3149	146-300	G-BTZN	19/12/89	17/04/97	BAE Asset Mgmt	G-11-149/HS-TBN/G-BTZN / N146PZ/ZP-CCY/EI-CLY	W	Std Filton 04/08/03
E3150	146-300QT	OO-TAK	16/08/89	01/01/00	TNT Airways	G-BRGK/SE-DIM/G-TJPM	H	
E3151	146-300QT	OO-TAA	24/09/89	01/01/00	TNT Airways	G-BRGM/SE-DIT/G-TNTR	H	
E1152	146-100	VH-NJR	27/04/90	19/02/92	Nat. Jet Systems	G-6-152/G-BRLN/PK-DTC	W	
E3153	146-300QT	OO-TAJ	14/12/89	01/01/00	TNT Airways	G-BRPW/G-TNTE	H	
E3154	146-300QT	OO-TAS	14/03/90	01/01/00	BAE Asset Mgmt	G-6-154/G-BRXI/G-TNTF/EC-712/EC-FFY	H	
E3155	146-300	G-BTNU	11/04/90	01/03/96	Air Wisconsin	G-6-155/G-BSLS/G-BTNU/EI-CLJ	W	
E2156	146-200	N156TR	14/12/89	30/12/96		G-11-156/N884DV	W	
E3157	146-300	EI-DEX	14/02/90	11/06/04	CityJet	G-6-157/G-UKID	W	
E3158	146-300	G-UKRC	18/05/90	01/04/00	BAE Asset Mgmt	G-6-158/G-BSMR	W	Std Norwich 08/01/04
E3159	146-300	EI-CLI	30/01/90	19/04/95	Aer Lingus	G-5-159/I-ATSD/G-BVSA	H	Std Mojave 03/07/04
E1160	146-100	VH-NJD	15/08/90	24/11/95	Nat Jet Systems	G-6-160/A2-ABF/G-BVLJ/EI-CJP/VH-JSF	W	Scr Woodford 09/01
E3161	146-300	B-1775	28/02/90	03/08/90	UNI AIR	G-6-161/G-BSOC/B-1775	H	Std Norwich 07/01/04
E3162	146-300	G-UKAG	12/11/90	01/04/00	BAE Asset Mgmt	G-6-162	W	1st production EFIS aircraft
E3163	146-300	D-AEWA	04/05/90	23/03/96	Eurowings	G-6-163/G-BTJG/EC-876/EC-FIU/ HB-IXY	W	
E2164	146-200	OO-DJE	20/06/90	15/02/02	Sn Brussels	G-6-164	H	
E3165	146-300	G-BSNR	12/05/90	01/04/00	BAE Asset Mgmt	G-6-165/EC-807/EC-FGT	H	Std Filton 22/08/03
E3166	146-300QT	OO-TAD	30/06/90	01/01/00	TNT Airways	G-6-166/G-BSLZ/RP-C480/G-TNTM	W	
E2167	146-200	OO-DJF	30/04/90	15/02/02	Sn Brussels	G-6-167	H	
E3168	146-300QT	OO-TAH	14/06/90	01/01/00	TNT Airways	G-BSGI/RP-C479/G-TNTL	W	
E3169	146-300	G-BSNS	01/06/90	20/05/04	Flightline	G-6-169/EC-839/EC-FHU/G-BSNS/EI-CTN	W	
E2170	146-200	VH-NJG	10/08/90	26/08/94	Nat. Jet Systems	G-BSOH/I-FLRX	H	
E2171	146-200	G-OINV	09/06/90	31/03/02	BA CitiExpress	G-6-171/VH-EWI	H	
E2172	146-200	OO-DJH	14/09/90	15/02/02	Sn Brussels	G-6-172/G-BSSG	W	
E3173	146-300	ZK-NZM	18/08/90	08/06/01	Mount Cook A/L	G-6-173/VH-EWJ	H	Std Melbourne 12/05/03
E3174	146-300	G-BSXZ	30/09/90	24/01/02	Flightline	G-6-174/B-1776/G-NJIB	W	
E3175	146-300	ZK-NZL	06/09/90	08/06/01	Mount Cook A/L	G-6-175/VH-EWK	H	Std Melbourne 21/12/01

MSN	Type	Current Registration	First flight	Delivery to current operator	Current/last operator	Previous registrations	Built	Notes
E2176	146-200QC	G-ZAPO	17/09/90	11/08/00	Titan Airways	G-PRCS/VH-NJQ/G-BWLG/F-GMMP	H	
E3177	146-300	ZK-NZN	28/09/90	08/06/01	Mount Cook A/L	G-6-177/VH-EWL	H	Std Melbourne 27/06/02
E3178	146-200	VH-NJH	21/12/90	25/08/94	Nat Jet Systems	G-6-178/G-BTCP/I-FLRW	W	
E3179	146-200	VH-EWM	21/10/90	31/10/93	Ansett Australia	G-6-179	H	Std Melbourne 04/07/02
E2180	146-200	OO-DJG	30/11/90	15/02/02	Sn Brussels	G-6-180	W	
E3181	146-300	G-JEBA	31/10/90	16/06/98	Flybe	G-6-181/G-BSYR/HS-TBL	H	
E3182	146-300QT	OO-TAE	15/11/90	01/01/00	TNT Airways	G-BSUY/G-TNTG	H	
E3183	146-300	D-AEWB	19/10/90	14/08/96	Eurowings	G-6-183/G-BSYS/G-BUHB	W	
E3184	146-300	VH-NJJ	11/02/91	26/08/94	Nat Jet Systems	G-6-184/G-BKTC/I-FLRV	W	
E3185	146-300	G-JEBB	13/12/90	30/06/98	Flybe	G-6-185/HS-TBK	W	
E3186	146-300QT	OO-TAF	14/12/90	01/01/00	TNT Airways	G-6-186/G-BSXL/G-TNTK	H	
E3187	146-300	D-AHOI	30/11/90	18/12/97	Eurowings	G-6-187/G-BSYT/EC-899/EC-FKF	H	
E2188	146-200QC	F-GLNI	20/02/91	07/10/91	Axis Airways	G-6-188/G-BTDO	W	
E3189	146-300	G-JEBC	22/12/90	04/06/98	Flybe	G-6-189/HS-TBO	H	
E3190	146-300	ZK-NZK	07/03/91	08/06/01	Mount Cook A/L	G-6-190/VH-EWN	H	Std Melbourne 08/04/03
E3191	146-300	G-JEBD	19/04/91	31/07/98	Flybe	G-6-191/HS-TBJ	H	
E2192	146-200	OO-MJE	10/03/91	15/02/02	Sn Brussels	G-6-192	W	
E3193	146-300	G-BUHC	05/07/91	30/05/00	Flightline	G-6-193/G-BTMI/G-BUHC/EI-CTO	W	
E3194	146-300QT	VH-NJM	10/04/91	28/11/95	Nat Jet Systems	G-6-194/G-BTHT/N599MP	H	Std Melbourne 05/08/02
E3195	146-300	VH-EWR	23/05/91	31/10/93	Ansett Australia	G-6-195	H	
E2196	146-200	SE-DRM	13/05/91	15/02/02	Malmö Aviation	G-6-196/OO-DJJ	W	
E3197	146-300	ZA-MEV	27/06/91	15/02/02	Albanian A/L	G-6-197/VH-EWS	H	
E3198	146-300QT	VH-NJS	20/06/91	06/10/92	Nat Jet Systems	G-6-198/G-BTLD	H	
E1199	146-100	A5-RGE	18/06/91	22/12/92	Druk Air	G-RJET/N170RJ	H	
E2200	146-200	D-ACFA	02/09/91	20/10/94	Eurowings	G-6-200/G-BTVT	W	
E2201	146-200	D-AJET	24/10/91	02/04/95	Eurowings	G-6-201	W	
E3202	146-300	G-BTUY	08/07/91	12/07/03	Buzz	G-6-202/G-NJIC	W	
E3203	146-300	G-BTTP	25/10/91	01/04/00	Buzz	G-6-203	H	Std Filton 06/1-/03
E2204	146-200	I-FLRU	21/02/92	27/05/04	Club Air	G-6-204/G-OSAS/I-FLRA	W	
E3205	146-300	G-BTVO	02/10/91	23/10/04	Flightline	G-6-205/B-1777/G-NJID	W	
E3206	146-300	G-JEBE	17/12/91	28/05/98	Flybe	G-6-206/HS-TBM	W	
E3207	146-300	B-2711	19/02/92	02/12/92	China Eastern	G-6-207/G-BUHV	W	
E2208	RJ85	N501XJ	23/03/92	24/04/97	Mesaba Airlines	G-ISEE/G-6-208/G-ISEE	H	First RJ85 Last Hatfield aircraft
E3209	146-300	G-JEBG	06/03/92	15/07/03	Flybe	G-6-209/G-NJIE/G-BVCE	W	
E2210	146-200	G-BVMP	30/03/92	21/05/94	BAE Asset Mgmt	G-6-210/G-BVMP/I-FLRE	W	Std Kemble 19/05/04
E2211	146-200	F-GOMA	13/09/93	17/06/94	Axis Airways	G-6-211/G-BVCD	W	
E3212	146-300	B-2712	31/07/92	14/12/92	China Eastern	G-6-212	W	

E3213	146-300	VH-NJL	21/07/92	20/10/94	Nat.Jet Systems	G-6-213/G-BVPE	W	
E3214	146-300	B-2715	23/11/92	23/12/92	China Eastern	G-6-214	W	
E3215	146-300	B-2716	05/12/92	29/12/92	China North West	G-6-215	W	Cra Yinchuan 23/07/93
E3216	146-300	B-2717	18/02/93	26/03/93	China Eastern	G-6-216	W	
E3217	146-300	VH-NJN	23/12/91	27/10/94	Nat.Jet Systems	G-6-217/G-BUHW	W	
E3218	146-300	B-2719	24/03/93	22/04/93	China Eastern	G-6-21	W	
E3219	146-300	B-2720	05/04/93	25/08/93	China Eastern	G-6-219	W	
E2220	146-200	I-FLRI	02/02/93	01/07/94	Club Air	G-6-220/G-BVMT	W	
E3221	RJ100	SE-DSO	13/05/92	06/08/01	Malmö Aviation	G-OIII/N504MM	W	First RJ100
E3222	146-300	B-2718	24/06/93	12/01/94	China Eastern	G-6-222	W	
E1223	RJ70	YL-BAK	24/08/93	19/03/96	Air Baltic	G-6-223/N832BE/YL-BAK/VH-NJW/G-BZFA/EI-CUO	W	
E1224	RJ70	YL-BAL	23/09/93	23/04/96	Air Baltic	G-6-224/N833BE	W	
E1225	RJ70	YL-BAN	29/10/93	24/01/96	Air Baltic	G-6-225/N834BE	W	
E2226	RJ85	HB-IXF	27/11/92	31/03/02	Swiss	G-CROS	W	Std Kemble 06/04
E2227	146-200	G-BVMS	11/03/94	23/06/94	BAE Asset Mgnt	G-6-227/G-BVMS/I-FLRO	W	
E1228	RJ70	VH-NJT	21/01/94	08/12/95	Nat.Jet Systems	G-6-228/G-OLXX	W	First RJ70
E1229	RJ70	TC-THI	23/07/92	28/06/96	THY	G-BUFI	W	
E1230	RJ70	TC-THJ	15/03/96	29/03/96	THY	G-6-230/N835BE	W	
E2231	RJ85	HB-IXG	02/05/93	31/03/02	Swiss	G-6-231	W	
E2232	RJ100	TC-THA	12/07/93	22/07/93	THY	G-6-232	W	
E2233	RJ85	HB-IXH	02/05/93	31/03/02	Swiss	G-6-233/G-XARJ	W	
E3234	RJ100	G-CCTB	06/08/93	12/08/93	BAE Asset Mgnt	G-6-234/TC-THB	W	Std Exeter 22/09/04
E2235	RJ85	HB-IXK	18/06/93	31/03/02	Swiss	G-6-235	W	
E3236	RJ100	G-CDCN	01/09/93	07/09/93	BAE Asset Mgnt	G-6-236/TC-THC	W	Std Kemble 14/10/04
E3237	RJ100	TC-THD	08/10/93	16/11/93	THY	G-6-237	W	
E3238	RJ100	TC-THE	26/10/93	08/11/93	THY	G-6-238	W	
E2239	RJ85	PK-PJJ	28/08/93	20/12/93	Pelita Air	G-BVAE	W	
E3240	RJ100	TC-THF	04/02/94	09/03/94	THY	G-6-240	W	Dbr Samsum 11/01/98
E3241	RJ100	TC-THG	04/03/94	23/03/94	THY	G-6-241	W	Cr Diyarbakir 08/01/03
E3242	RJ100	SE-DSP	11/06/94	13/03/01	Malmö Aviation	G-6-242/N505MM	W	
E3243	RJ100	TH-THH	30/03/94	14/04/94	THY	G-6-243	W	
E3244	RJ100	SE-DSR	14/07/94	26/04/01	Malmö Aviation	G-6-244/N506MM	W	
E3245	RJ100	SE-DSS	05/08/94	25/05/01	Malmö Aviation	G-6-245/N507MM	W	
E2246	RJ85	D-AVRO	11/10/94	18/10/94	Lufthansa Cityline	G-6-246	W	
E3247	RJ100	SE-DST	17/06/94	14/11/00	Malmö Aviation	G-6-247/N508MM	W	
E3248	RJ100	SE-DSU	30/08/94	03/04/01	Malmö Aviation	G-6-248/N509MM	W	
E1249	RJ70	TC-THL	11/06/96	24/06/96	THY	G-6-249	W	Dbr Siirt 22/04/00
E3250	RJ100	SE-DSV	18/09/94	15/12/00	Malmö Aviation	G-6-250/N510MM	W	
E2251	RJ85	D-AVRC	22/02/95	03/03/95	Lufthansa Cityline	G-6-251	W	
E1252	RJ70	TC-THN	24/04/96	03/05/96	THY	G-6-252	W	
E2253	RJ85	D-AVRB	01/12/94	09/12/94	Lufthansa Cityline	G-6-253/G-BVWD	W	
E1254	RJ70ER	EI-COQ	25/08/94	28/10/97	Azzurair	G-6-254/G-BVRJ/9H-ACM	W	

MSN	Type	Current Registration	First flight	Delivery to current operator	Current/last operator	Previous registrations	Built	Notes
E3255	RJ100	SE-DSX	16/11/94	01/08/01	Malmö Aviation	G-6-255/N511MM	W	
E2256	RJ85	D-AVRA	10/11/94	18/11/94	Lufthansa Cityline	G-6-256	W	
E2257	RJ85	D-AVRD	14/03/95	17/03/95	Lufthansa Cityline	G-6-257	W	
E1258	RJ70ER	EI-CPJ	02/02/05	20/03/98	EuroManx	G-6-258/9H-ACN	W	
E3259	RJ100	HB-IXT	30/03/95	31/03/02	Swiss	G-BVYS	W	
E1250	RJ70ER	EI-CPK	10/12/94	23/03/98	Azzurair	G-6-260/9H-ACO	W	
E2261	RJ85	D-AVRE	24/03/95	31/03/95	Lufthansa Cityline	G-6-261	W	
E3262	RJ100	HB-IXX	03/10/95	31/03/02	Swiss	G-6-262	W	
E3263	RJ100	SE-DSY	19/07/95	15/02/01	Malmö Aviation	G-6-263/N512MM	W	
E3264	RJ100	TC-THM	13/06/95	22/06/95	THY	G-6-264	W	
E3265	RJ100	TC-THO	22/06/95	28/06/95	THY	G-6-265	W	
E2266	RJ85	D-AVRG	02/08/95	08/09/95	Lufthansa Cityline	G-6-266	W	
E1267	RJ70ER	EI-CPL	08/03/95	06/03/98	Azzurair	G-6-267/9H-ACP	W	
E2268	RJ85	D-AVRH	20/08/95	12/10/95	Lufthansa Cityline	G-6-268/G-OCLH	W	
E2269	RJ85	D-AVRF	16/05/95	30/06/95	Lufthansa Cityline	G-JAYV	W	
E2270	RJ85	D-AVRI	02/11/95	07/12/95	Lufthansa Cityline	G-6-270/G-CLHX	W	
E2271	RJ85	OO-DJK	17/11/95	15/02/02	SN Brussels	G-6-271	W	
E3272	RJ100	HB-IXW	20/10/95	31/03/02	Swiss	G-6-272	W	
E2273	RJ85	OO-DJL	23/11/95	15/02/02	SN Brussels	G-6-273	W	
E3274	RJ100	HB-IXV	18/11/95	01/07/02	Swiss	G-6-274	W	
E2275	RJ85	OO-DJN	07/12/95	15/02/02	SN Brussels	G-6-275	W	
E3276	RJ100	HB-IXU	09/12/95	01/07/02	Swiss	G-6-276	W	
E2277	RJ85	D-AVRJ	26/01/96	22/02/96	Lufthansa Cityline	G-6-277/G-BWKY	W	
E2278	RJ85	D-AVRK	04/03/96	21/03/96	Lufthansa Cityline	G-6-278	W	
E2279	RJ85	OO-DJO	14/12/95	15/02/02	SN Brussels	G-6-279	W	
E3280	RJ100	HB-IXS	03/02/96	01/07/02	Swiss	G-6-280	W	
E3281	RJ100	HB-IXR	31/03/02	01/07/02	Swiss	G-6-281	W	
E3282	RJ100	HB-IXQ	16/03/96	01/07/02	Swiss	G-6-282	W	
E3283	RJ100	HB-IXP	16/04/96	01/07/02	Swiss	G-6-283	W	
E3284	RJ100	HB-IXO	22/05/96	01/07/02	Swiss	G-6-284	W	
E2285	RJ85	D-AVRL	24/03/96	28/03/96	Lufthansa Cityline	G-6-285	W	
E3286	RJ100	HB-IXN	14/07/96	31/03/02	Swiss	G-6-286	W	
E2287	RJ85	OO-DJP	28/04/96	15/02/02	SN Brussels	G-6-287	W	
E2288	RJ85	D-AVRM	18/05/96	24/05/96	Lufthansa Cityline	G-6-288	W	
E2289	RJ85	OO-DJQ	19/06/96	15/02/02	SN Brussels	G-6-289	W	
E2290	RJ85	OO-DJR	19/07/96	15/02/02	SN Brussels	G-6-290	W	
E3291	RJ100	HB-IXM	16/08/96	22/08/96	Crossair	G-6-291	W	Cra Nr Zurich 24/11/01
E2292	RJ85	OO-DJS	13/08/96	15/02/02	SN Brussels	G-6-292	W	

E2293	RJ85	D-AVRN	04/09/96	12/09/96	Lufthansa Cityline	G-6-293	W
E2294	RJ85	OO-DJT	19/09/96	15/02/02	SN Brussels	G-6-294	W
E2295	RJ85	OO-DJV	03/10/96	15/02/02	SN Brussels	G-6-295	W
E2296	RJ85	OO-DJW	15/10/96	15/02/02	SN Brussels	G-6-296	W
E2297	RJ85	OO-DJX	09/11/96	15/02/02	SN Brussels	G-6-297	W
E3298	RJ100	G-BXAR	15/03/97	31/03/97	BA CitiExpress	G-6-298	W
E2299	RJ85	EI-CNI	17/11/96	26/11/96	Azzuraair	G-6-299	W
E2300	RJ85	EI-CNJ	01/12/96	06/12/96	Azzuraair	G-6-300	W
E3301	RJ100	G-BXAS	15/04/97	01/05/97	BA CitiExpress	G-6-301	W
E2302	RJ85	OO-DJY	29/01/97	15/02/02	SN Brussels	G-6-302	W
E2303	RJ85	D-AVRP	09/02/97	21/02/97	Lufthansa Cityline	G-6-303	W
E2304	RJ85	D-AVRQ	12/03/97	21/03/97	Lufthansa Cityline	G-6-304	W
E2305	RJ85	OO-DJZ	04/04/97	15/02/02	SN Brussels	G-6-305	W
E2306	RJ85	A9C-HWR	03/05/97	08/05/97	Bahrain Defence	G-6-306/EI-CNK	W
E2307	RJ85	N502XJ	20/05/97	22/05/97	Mesaba Airlines	G-6-307	W
E3308	RJ100	OO-DWA	13/06/97	15/02/02	SN Brussels	G-BXEU/G-6-308	W
E2309	RJ85	UK-80002	19/06/97	04/07/97	Uzbekistan A/w	G-6-309	W
E2310	RJ85	N503XJ	25/06/97	30/06/97	Mesaba Airlines	G-6-310	W
E2311	RJ85	N504XJ	17/07/97	24/07/97	Mesaba Airlines	G-6-311	W
E2312	RJ85	UK-80001	12/11/97	18/12/97	Uzbekistan A/w	G-6-312	W
E2313	RJ85	N505XJ	20/08/97	25/08/97	Mesaba Airlines	G-6-313	W
E2314	RJ85	N506XJ	13/09/97	17/09/97	Mesaba Airlines	G-6-314	W
E3315	RJ100	OO-DWB	13/06/97	15/02/02	SN Brussels	G-6-315	W
E2316	RJ85	N507XJ	17/10/97	23/10/97	Mesaba Airlines	G-6-316	W
E2317	RJ85	D-AVRR	11/11/97	09/12/97	Lufthansa Cityline	G-6-317	W
E2318	RJ85	N508XJ	02/12/97	02/12/97	Mesaba Airlines	G-6-318	W
E2319	RJ85	UK-80003	13/12/97	24/12/97	Uzbekistan A/w	G-6-319	W
E3320	RJ100	G-BZAT	21/12/97	10/01/98	Cityflyer Express	G-6-320	W
E2321	RJ85	N509XJ	20/01/98	28/01/98	Mesaba Airlines	G-6-321	W
E3322	RJ100	OO-DWC	14/02/98	15/02/02	SN Brussels	G-6-322	W
E2323	RJ85	N510XJ	02/03/98	09/03/98	Mesaba Airlines	G-6-323	W
E3324	RJ100	OO-DWD	25/03/98	15/02/02	SN Brussels	G-6-324	W
E2325	RJ85	N511XJ	03/04/98	08/04/98	Mesaba Airlines	G-6-325	W
E2326	RJ85	N512XJ	21/04/98	24/04/98	Mesaba Airlines	G-6-326	W
E3327	RJ100	OO-DWE	19/05/98	15/02/02	SN Brussels	G-6-327	W
E3328	RJ100	G-BZAU	03/06/98	08/06/98	Cityflyer Express	G-6-328	W
E2329	RJ85	N513XJ	13/06/98	22/06/98	Mesaba Airlines	G-6-329	W
E2330	RJ85	N514XJ	24/06/98	29/06/98	Mesaba Airlines	G-6-330	W
E3331	RJ100	G-BZAV	16/07/98	25/07/98	Cityflyer Express	G-6-331	W
E3332	RJ100	OO-DWF	05/08/98	15/02/02	SN Brussels	G-6-332	W
E2333	RJ85	N515XJ	12/08/98	18/08/98	Mesaba Airlines	G-6-333	W
E2334	RJ85	N516XJ	02/09/98	21/09/98	Mesaba Airlines	G-6-334	W

MSN	Type	Current Registration	First flight	Delivery to current operator	Current/last operator	Previous registrations	Built	Notes
E2335	RJ85	N517XJ	26/09/98	01/10/98	Mesaba Airlines	G-6-335	W	
E3336	RJ100	OO-DWG	08/10/99	15/02/02	SN Brussels	G-6-336	W	
E2337	RJ85	N518XJ	20/10/98	29/10/98	Mesaba Airlines	G-6-337	W	
E3338	RJ100	HB-IYZ	05/11/98	31/03/02	Swiss	G-6-338	W	
E3339	RJ100	HB-IYY	01/12/98	31/03/02	Swiss	G-6-339	W	
E3340	RJ100	OO-DWH	08/12/98	15/02/02	SN Brussels	G-6-340	W	
E3341	RJ100	SX-DVA	20/12/98	28/10/01	Aegean Airlines	G-6-341/TC-TRA	W	
E3342	RJ100	OO-DWI	15/01/99	15/02/02	SN Brussels	G-6-342	W	
E3343	RJ100	SX-DVB	09/04/99	28/10/01	Aegean Airlines	G-6-343	W	
E2344	RJ85	N519XJ	04/02/99	18/02/99	Mesaba Airlines	G-6-344	W	
E2345	RJ85	N520XJ	19/02/99	25/02/99	Mesaba Airlines	G-6-345	W	
E2346	RJ85	N521XJ	03/03/99	10/03/99	Mesaba Airlines	G-6-346	W	
E2347	RJ85	N522XJ	15/03/99	24/03/99	Mesaba Airlines	G-6-347	W	
E2348	RJ85	N523XJ	26/03/99	01/04/99	Mesaba Airlines	G-6-348	W	
E2349	RJ85	N524XJ	21/04/99	27/04/99	Mesaba Airlines	G-6-349	W	
E2350	RJ85	N525XJ	04/05/99	16/05/99	Mesaba Airlines	G-6-350	W	
E2351	RJ85	N526XJ	17/05/99	25/05/99	Mesaba Airlines	G-6-351	W	
E2352	RJ85	N527XJ	31/05/99	08/06/99	Mesaba Airlines	G-6-352	W	
E2353	RJ85	N528XJ	19/06/99	28/06/99	Mesaba Airlines	G-6-353	W	Dbr 15/10/02 Scr 09/04
E2354	RJ100	G-BZAW	02/07/99	16/07/99	BA CitiExpress	G-6-354	W	
E3355	RJ100	OO-DWJ	22/07/99	15/02/02	SN Brussels	G-6-355	W	
E3356	RJ100	G-BZAX	06/08/99	19/08/99	BA CitiExpress	G-6-356	W	
E3357	RJ100	HB-IYX	28/08/99	31/03/02	Swiss	G-6-357	W	
E3358	RJ100	SX-DVC	12/09/99	28/10/01	Aegean Airlines	G-6-358	W	
E3359	RJ100	HB-IYW	30/09/99	31/03/02	Swiss	G-6-359	W	
E3360	RJ100	OO-DWK	07/10/99	15/02/02	SN Brussels	G-6-360	W	
E3361	RJ100	OO-DWL	22/10/99	15/02/02	SN Brussels	G-6-361	W	
E3362	RJ100	SX-DVD	16/11/99	28/10/01	Aegean Airlines	G-6-362	W	
E2363	RJ85	N529XJ	01/12/99	07/12/99	Mesaba Airlines	G-6-363	W	
E2364	RJ85	N530XJ	20/12/99	12/01/00	Mesaba Airlines	G-6-364	W	
E2365	RJ85	N531XJ	17/01/00	25/01/00	Mesaba Airlines	G-6-365	W	
E2366	RJ85	N532XJ	10/02/00	01/03/00	Mesaba Airlines	G-6-366	W	
E2367	RJ85	N533XJ	25/02/00	08/03/00	Mesaba Airlines	G-6-367	W	
E3368	RJ100	G-BZAY	14/03/00	30/03/00	BA CitiExpress	G-6-368	W	
E3369	RJ100	G-BZAZ	31/03/00	31/07/02	BA Citiexpress	G-6-369	W	
E2370	RJ85	N534XJ	14/04/00	27/04/00	Mesaba Airlines	G-6-370	W	
E2371	RJ85	N535XJ	02/05/00	09/05/00	Mesaba Airlines	G-6-371	W	
E2372	RJ85	N536XJ	18/05/00	26/05/00	Mesaba Airlines	G-6-372	W	

E3373	RJ100	G-CFAA	07/06/00	16/06/00	BA CitiExpress	G-6-373	W
E3374	RJ100	SX-DVE	21/06/00	28/10/01	Aegean Airlines	G-6-374	W
E3375	RJ100	SX-DVF	06/07/00	28/10/01	Aegean Airlines	G-6-375	W
E2376	RJX85	G-ORJX	28/04/01		BAE Systems		W 1st RJX85 Final flight 10/01/02 Std Woodford
E3377	RJ100	G-CFAB	19/10/00	30/11/00	BA CitiExpress	G-6-377	W
E3378	RJX100	G-IRJX	23/09/01		BAE Systems		W 1st RJX100. Final flight to Manchester Airport 6/2/03
E3379	RJ100	G-CFAC	06/12/00	15/12/00	BA CitiExpress	G-6-379	W
E3380	RJ100	G-CFAD	12/01/01	26/01/01	BA CitiExpress	G-6-380	W
E3381	RJ100	G-CFAE	13/02/01	26/02/01	BA CitiExpress	G-6-381	W
E3382	RJ100	G-CFAF	15/03/01	23/03/01	BA CitiExpress	G-6-382	W
E2383	RJ85	OH-SAH	24/04/01	10/05/01	Blue 1	G-6-383	W
E3384	RJ100	G-CFAH	15/05/01	08/06/01	BA CitiExpress	G-6-384	W
E2385	RJ85	OH-SAI	11/06/01	20/06/01	Blue 1	G-6-385	W
E3386	RJ100	OH-SAM	28/08/01	22/10/03	Blue 1	G-6-386/G-NBAA	W
E3387	RJ100	OH-SAN	25/02/02	28/10/03	Blue 1	G-6-387/G-CBMF	W
E2388	RJ85	OH-SAJ	24/07/01	15/08/01	Blue 1	G-6-388	W
E2389	RJ85	OH-SAK	14/09/01	27/09/01	Blue 1	G-6-389	W
E2390	RJ85	A9C-BDF	03/10/01	09/11/01	Bahrain Def Force	G-6-390	W
E3391	RJX100	G-6-391	09/01/02		BAE Systems	G-6-391	W 1st production RJX Bu Woodford 07/04
E2392	RJ85	OH-SAL	20/11/01	29/11/01	Blue 1	G-6-392	W
E2393	RJ85	OH-SAO	22/03/02	04/11/03	Blue 1	G-6-393/G-CMBG	W
E2394	RJ85	OH-SAP	26/04/02	26/11/03	Blue 1	G-6-394/G-CMBH	W Last British airliner delivery

Appendix 3
Chronology of the BAe146/RJ/RJX

January 1960	De Havilland becomes part of the Hawker Siddeley Group. Studies had already taken place on a small feederliner and continued through the 1960s.
1963	Hawker Siddeley Aviation (HSA) formed.
29 August 1973	HSA announces the go-ahead of the HS 146.
21 October 1974	HSA halts work on HS 146 owing to world economic situation.
9 December 1974	HSA agrees to maintain the jigs, tools, drawings and design capacity as the Government wanted to maintain this civil airliner capability when the aerospace industry was nationalised.
29 April 1977	Nationalised British Aerospace formed from British Aircraft Corporation, Hawker Siddeley Aviation and Scottish Aviation.
10 July 1978	Government announced the go-ahead of the BAe 146.
January 1981	British Aerospace privatised.
20 May 1981	BAe 146 rolled out at Hatfield (E1001).
3 September 1982	First flight of BAe 146-100 (E1001).
1 August 1982	First flight of BAe 146-200 (E2008).
4 February 1983	Certification of BAe 146-100.
23 May 1983	First delivery of a BAe 146-100 (E1006) to Dan-Air.
27 May 1983	Dan-Air operates the first 146 service from Gatwick to Dublin.
3 June 1983	Certification of BAe 146-200.
16 June 1983	First delivery of a BAe 146-200 (E2012) to Air Wisconsin.
21 August 1986	First flight of BAe 146-200QT (E2056) after conversion at Dothan, Alabama.
1 May 1987	BAe 146-300 (E3001) first flight at Hatfield after conversion from 100 series.
16 May 1988	First flight of the first BAe 146 (E2106) assembled at Woodford (all previous 146s were built at Hatfield).
6 September 1988	Certification of BAe 146-300.
16 December 1988	First delivery of a BAe 146-300 (E3120 to Air Wisconsin).
29 May 1989	First flight of BAe146-200QC, (E2119) after conversion at Dothan, Alabama.
March 1991	BAe announces that 146 final assembly at Hatfield is to end.
23 March 1992	First flight of the final Hatfield assembled aircraft RJ85 (E2208).
13 May 1992	First flight of RJ100 (E3221) at Woodford.
June 1992	RJ (Regional Jet) branding introduced for developed 146. RJ70 (146-100), RJ85 (146-200), and RJ100 (146-300).
23 July 1992	First flight of Avro RJ70 (E1229) at Woodford.
23 September 1992	BAe announces the closure of Hatfield.
January 1993	BAe establishes the Asset Management Organisation at Hatfield.
23 April 1993	Certification of Avro RJ85.
23 April 1993	First delivery of an Avro RJ85 (E2226 to Crossair).
2 July 1993	Certification of Avro RJ100.
22 July 1993	First delivery of an Avro RJ100 (E3232 to THY).

Appendices

24 August 1993	Certification of Avro RJ70.
11 September 1993	First delivery of an Avro RJ70 (E1223 to Business Express).
4 April 1994	Hatfield airfield closed.
2 January 1996	Regional aircraft joint venture with Aerospatiale and Alenia, Aero International (Regional) established.
April 1998	Aero International (Regional) dissolved.
November 1999	BAE Systems formed by the merger of British Aerospace and Marconi Electronic Systems.
21 March 2000	BAE Systems formally launches the Avro RJX.
30 April 2001	First flight of Avro RJX85 (E2376) at Woodford.
23 September 2001	First flight of Avro RJX100 (E3378) at Woodford.
27 November 2001	BAE Systems ends the Avro RJX programme.
26 April 2002	Very last 'first flight' of a British-built commercial airliner when RJ85 G-CBMH (E2394) makes its maiden flight at Woodford.
26 April 2002	BAE Systems Regional Aircraft announces it is to relaunch as a service business following the cessation of Regional Aircraft manufacturing by the company.
1 October 2003	First flight of G-LUXE (E3001) as an Atmospheric Research Aircraft from Woodford.
26 November 2003	Final delivery of a newly built British airliner – RJ85 (E2394) to Blue 1.
10 May 2004	Delivery of G-LUXE (E3001) as an Atmospheric Research Aircraft.

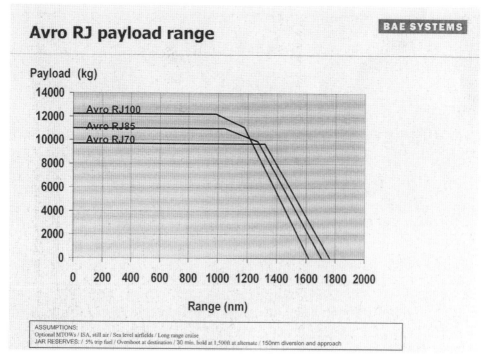

Payload Range Curve for the Avro RJ70, RJ85 and RJ100.

Appendix 4
Aircraft Data BAe 146 and RJ

	BAe 146-100	BAe 146-200	BAe 146-30
Maximum Take Off-Weight	38,102kg (84000lb)	42,184kg (92,000lb)	43,091kg (95,000lb)
Maximum Landing Weight	35,153kg (77,500lb)	36,741kg (81,000lb)	37,648kg (83000lb)
Maximum Zero Fuel Weight	31,070kg (68500lb)	33,340kg (73,500lb)	35,607kg (78,500lb)
Operating Empty Weight (typical)	23,290kg (51,342lb)	23,754kg (52,368lb	24,174kg (54,481lb)
Seats (5 abreast)	70	85	100
Seats (6 abreast)	82	100	112
Wing Span	26.34m (86ft 55in)	26.34m (86ft 55in)	26.34m (86ft 55in)
Length	26.16m (85ft 10in)	28.55m (93ft 8in)	30.1m (101ft 8in)
Cabin Length	15.42m (50 t 7in)	17.81m (58ft 5in)	20.20m (66ft 3in)
Seat Width (5 abreast)	48.25cm (19in)	48.25cm (19in)	48.25cm (19in)
Aisle Width (5 abreast)	53.3cm (21in)	53.3cm (21in)	53.3cm (21in)
Cabin Headroom	2.07m (6ft 9.in)	2.07m (6ft 9in)	2.07m (6ft 9in)
Freight Holds	13.56m3 (479 cuft)	18.25m3 (645 cuft)	22.98m (812 cuft)
Max Speed	m.073/300kt IAS	m.073/300kt IAS	m.073/300kt IAS
Max Altitude	31,000ft	31,000ft	31,000ft
Powerplant Avco Lycoming	ALF502 R-3 (6,700lb) or ALF502R-3A or ALF502R-5 (6,970lb)	As 146-100	ALF502 R-5 (7,000lb) or LF507- 1H(7,000lb)

OH-SAO (E2393), an RJ85 delivered to Blue 1 in November 2003. (Ian Lowe)

	RJ70	**RJ85**	**RJ100**
Seats (5 abreast)★	70	85	100
Seats (6 abreast)★	82	100	112
Wing Span	26.34m (86ft 55in)	26.34m (86ft 55in)	26.34m (86ft 55in)
Length	26.16m (85ft 10in)	28.55m (93ft 8in)	30.1m (101ft 8in)
Cabin Length	15.42m (50 t 7in)	17.81m (58ft 5in)	20.20m (66ft 3in)
Seat Width (5 abreast)	48.25 cm (19in)	48.25 cm (19in)	48.25cm (19in)
Aisle Width (5 abreast)	53.3 cm (21in)	53.3 cm (21in)	53.3cm (21in)
Cabin Headroom	2.07m (6ft 9.in)	2.07m (6ft 9.in)	2.07m (6ft 9in)
Freight Holds	13.56m3 (479 cuft)	1,8.25m3 (645 cuft)	22.98m (812 cuft)
Runway for 740 km	1,091m (3579ft)	1,157m (3796ft)	1314m (4,311ft)
Range	2,998 km (1619 nm)	2,796 km (1510 nm)	2554 km (1,379 nm)
Max Speed	m.073/300kt IAS	m.073/300kt IAS	m.073/300kt IAS
Max Altitude	35,000ft	35,000ft	35,000ft
Powerplant	LF507-1F (7,000lb)	LF507-1F (7,000lb)	LF507-1F (7,000lb)

Textron Lycoming
★ seat pitch 33in

Bibliography

Interviews
Peter Sedgwick, former Chief Test Pilot, 25 November 2002
Roger de Mercado, former Chief Flight Test Engineer, 26 March 2004
Crew members of No.32 (The Royal) Squadron, Northolt, 2 April 2004
Maurice James, former Flight Test Engineer, 8 April 2004
Steve Doughty, Vice-President Sales & Marketing BAE Regional Aircraft, 12 May 2004
Dave Dorman, Head of Marketing & External Relations BAE Regional Aircraft, 12 May 2004
John Martin, former Chief Designer, 8 July 2004
Johnnie Johnstone, former Director of Marketing – Civil, 28 July 2004
Brian Botting, former Executive Director of Marketing – Civil, 28 July 2004
John Loader, former General Sales Manager – 146, 28 July 2004
Ken Pye, former Design Engineer, 9 November 2004
John Payne, former Project Manager Corporate, 10 November 2004
Dan Gurney, former Senior Project Pilot 146/RJ, 13 November 2004

Correspondence
Derek Ferguson, Flight Test Engineer BAE Systems Woodford
Ken Haynes, BAE Systems Woodford
Chris Grainger, Pilot, CityJet

Books
Against the Tide: Diaries, 1973-77, Tony Benn Arrow, 1990
BAe 146, M.J. Hardy, Ian Allan, 1991
Collision Course, Raymond Lygo, Book Guild, 2002
Hatfield Aerodrome – A History, Philip Birtles, BAe, 1993
Jet Airliner Production List 2004, TAHS 2004
The Northolt Story, I. Simpson, J. Willcox Crown Copyright 1999
The Spirit of Dan-Air, G.M. Simons, GMS 1993
TASS – The 146 Campaign, TASS 1991

Reports
Test Flying the 146 Mike Goodfellow Aerospace RAeS 1985

BAE Systems
BAe146/RJ brochures
BAe Hatfield & Woodford Newsletters

Journals
Aerospace
Aircraft Economics
Aircraft Illustrated
Air International
Air Pictorial
Airliner World
Flight International
Take Off
Welwyn & Hatfield Times

Web
www.airliners.net
www.regional-services.com
www.smiliner.com

Video
BAE 146 Flight on Film

Index

Abu Dhabi Government, 93,177
Aegean Airlines, 136-7,184-5
Aer Lingus, 62,63, 178-9
Aero International (Regional), 150,170,187
Aerosur, 77,91
Air Atlantic, 75,77
Air Baltic, 81,136,139,181
Air BC, 75
Air Botnia, 136-7,156
Air Botswana, 141,145,178
Air Cal, 63,65,71,73-4,162
Air Canada Jazz, 75,177-9
Air Dolomiti, 134
Air Malta, 29,130,133,135
Air New Zealand,7 9
Air Nova, 64,75
Air Pac, 67
Air UK, 60-1,63-4
Air Wisconsin, 22,35,41,46-7,56, 65-9,75,91,96, 139,161,165,176-9,186
Air Zimbabwe, 92-3,145,177
Airbus, 10,13,19,48,52,63-4,81,122,152,157, 170
Albanian Airlines, 81,162,176,177,180
Alitalia, 117,131,133
Allied Signal, See Avco Lcoming
American Airlines, 73-4,132
Ansett, 49,78-9,98,101,162,176-8,180
Ansett New Zealand, 79
Aspen Airways, 32,65,68,91,126
Asset Management Organisation, 10,63,75,121,1 33,159,161,169,185,
Atlantic Airways, 64,176-7
Australian Air Express, 78,99,112
Austrian Government, 111-2
Avco Lycoming, 14,24,27-8,52-3,70,83-4, 125-6,151-3,156-7,159,188
Aviacsa, 77-8
Axis Airways,99,180
Azzurra Air, 117,131,135
BAC One-Eleven, 12-13,19,20-1,41,59,135

Bahrain Defence Force, 140,183
Benn,Tony, 18-9,190
Beswick, Lord, 19,20
Blue 1, 9,135,185,187
Boeing, 13,19,21-2,35,48,69,70,79,98,122-3,131,133,137,149,151-2,157,161,170,172
Bombardier, 35,63,75,109,125,151,157,169-70
British Air Ferries, 41,46-7,50
British Airways,16,19,48,60,62,89,93,117,131,13 4,136,144,152,175-7,179,183-5
British Airways CitiExpress. See British Airways
British European. See Flybe
Bromma, 116,119,138
Business Express, 136,139,140,187
Buzz, 60-1,180
CAA. See Civil Aviation Authority.
CAAC, 80-1,84
CityJet, 7,63,117,177-9,190
Civil Aviation Authority, 40,43,46,49,115,129,173
Club Air, 77,163,175,177,181
Conti-Flug, 116
Crossair, 57-8,116-7,126,129,131-3,135,145,151,182,186
Dan-Air, 44,47,59,60,63,91,144,186,190
de Havilland, 7,9,11-13,66,85,113-4,116,122,174,186
Debonair, 61-2,64,68,73,75
DH. See de Havilland.
Druk Air, 81,136,152,158,159,177,180
East-West Airlines, 79
Embraer, 35,75,117,125,151,159,169-70
Eurowings, 63-4,134,176-9
FAAM. See Facility for Airborne Measurements.
Facility for Airborne Atmospheric Measurements, 9,149,165-6,175
Fairchild Dornier, 151,159,170
Filton, 9,10,23,36,60,83,137,175,178-80
Flightline, 62,71,73,117,175-6,178-80
Flybe, 9,62-3,81,117,137,143,147,153,155,157,1 59-60,175-80

Fokker, 11,14-16,20,22,35,51,78-9,117, 125-7,131-4,150,169-70
Formula One Administration, 91
Foster, Alan, 57,154-5,166
Goodfellow,Mike, 37-9,46,190
Grigg, Bob, 14-15,169
Gurney,Dan, 7,46,50,68,70,76,78,84,89,94,114, 122,128,146,188
Hall, Sir Arnold, 15-16,19
Hatfield, 6-7,9,11-15,17-20,23-4,33,36-40,46- 51,53-60,65-66,68,70,74,76-78,80-1,87,93- 4,96-99,101,109-110,114,120-4,128,135,138, 141,161,170-1,173-80,183-7,190
Hawker Siddeley, 7,10-20,22-3,41,83,85,185
Hayes International, 96,109
Hayman, Helen, 18
Honeywell. See Avco Lycoming.
HSA. See Hawker Siddeley.
Islay, 87-8,141
JAA. See Joint Airworthiness Authority.
Johnstone, Johnnie, 6,14,22,51,66,190
Joint Airworthiness Authority, 46,115,129,151,154,
Lan Chile, 77,143,176
Lockheed, 33,48,105,107,112,165
Loganair, 92,115,136
London City Airport, 7,63,89,113-5,117-9,125, 132,136,156,163,169,173,
Lufthansa, 62,64,117,119,121,131,133- 5,151,181-3
Lygo, Sir Raymond, 20,56-7,101,190
Makung International, 83
Mali Government, 87.92
Malmö Aviation, 117,138,178,181
Masefield, Charles, 53,55,122-3
McDonnell Douglas, 19,22,69,83,98,151,170
Mesaba. See Northwest Airlines
Minden Air Corporation, 163-5
Mistral Air, 99,177-8
Mojave Desert, 62-3,71-3,178-9
Moncrief Oil, 81,94,177

No.32 (The Royal) Squadron, 6,33,86,88- 90,94,141,190
Northwest Airlines, 123,138-9,145,180,183-4
Pauknair, 144
Pelita Air Services, 94,181
Presidential Airlines, 63,65,68-9,74-5,77
Prestwick, 9,23-4,66,161
Prince Charles, 87-9
Prince Philip, 49,85,87
Princess Diana, 89
PSA, 22,34,62-5,69-73,80,142,175
Qantas, 73,79,178,
Quantaslink. See Quantas
RAF, 6,49,85-91,94,107,111-2
Rolls-Royce,12-16,27,48,51,127
Ryanair, 61
Sabena, 117,121,131,134-5
SAM Colombia, 119,135,137-8,147
SAS, 131,136-7
Sedgwick, Peter, 7,37,39,47-9,56- 7,110,114,171,174,190
Small Tactical Airlifter. See STA
STA, 58,77,102,106-112,175
Swiss, 51,62,117,121,131-3,135,181-2,184
Swissair, See Swiss
TASS, 18-19,190
Textron Lycoming. See Avco Lycoming.
Thai Airways, 49,63,83
TNT, 79,96-101,176-80
Trident, Hawker Siddeley, 11-12,41,43,46
UNI AIR, 83,179
United Express, 58,65-6,68,74
US Army, 106,110
USAir, 62,71-3,132
Uzbekistan Airways, 94,140,183
WDL, 63,175-6,178
WestAir, 63,65,74
Woodford, 6-7,9-10,13,17,23,66,74,81,83,94,1 01,105,107,120-3,128-130,135,137,139,14 4,146,153-4,156-7,160-1,165,170,174,178- 85,190
Zimbabwean Governmental, see Air Zimbabwe

If you are interested in purchasing other books published by Tempus, or in case you have difficulty finding any Tempus books in your local bookshop, you can also place orders directly through our website

www.tempus-publishing.com